Math Made Nice-n-Easy Book™

 # 7

In This Book

- **Trigonometric Identities & Equations**

- **Straight Lines**

- **Conic Sections**

"MATH MADE NICE-n-EASY # 7" is one in a series of books designed to make the learning of math interesting and fun. For help with additional math topics, see the complete series of *"MATH MADE NICE-n-EASY"* titles.

Based on U.S. Government Teaching Materials

Research & Education Association
61 Ethel Road West
Piscataway, New Jersey 08854

Dr. M. Fogiel, Director

MATH MADE NICE-N-EASY BOOKS™
BOOK #7

Printed in the United States of America

Library of Congress Control Number 2001086083

International Standard Book Number 0-87891-206-1

MATH MADE NICE-N-EASY is a trademark of
Research & Education Association, Piscataway, New Jersey 08854

WHAT "MATH MADE NICE-N-EASY" WILL DO FOR YOU

The "Math Made Nice-n-Easy" series simplifies the learning and use of math and lets you see that math is actually interesting and fun. This series of books is for people who have found math scary, but who nevertheless need some understanding of math without having to deal with the complexities found in most math textbooks.

The "Math Made Nice-n-Easy" series of books is useful for students and everyone who needs to acquire a basic understanding of one or more math topics. For this purpose, the series is divided into a number of books which deal with math in an easy-to-follow sequence beginning with basic arithmetic, and extending through pre-algebra, algebra, and calculus. Each topic is described in a way that makes learning and understanding easy.

Almost everyone needs to know at least some math at work, or in a course of study.

For example, almost all college entrance tests and professional exams require solving math problems. Also, almost all occupations (waiters, sales clerks, office people) and all crafts (carpentry, plumbing, electrical) require some ability in math problem solving.

The "Math Made Nice-n-Easy" series helps the reader grasp quickly the fundamentals that are needed in using math. The reader is led by the hand, step-by-step, through the various concepts and how they are used.

By acquiring the ability to use math, the reader is encouraged to further his/her skills and to forget about any initial math fears.

The "Math Made Nice-n-Easy" series includes material originated by U.S. Government research and educational efforts. The research was aimed at devising tutoring and teaching methods for educating government personnel lacking a technical and/or mathematical background. Thanks for these efforts are due to the U.S. Bureau of Naval Personnel Training.

Dr. Max Fogiel
Program Director

Contents

Chapter 10
Conic Sections

Chapter 8
Trigonometric Identities & Equations

This is the final chapter in the section dealing directly with trigonometry and trigonometric relationships. This chapter includes the basic identities, formulas for additional identities involving multiples of an angle, and formulas for identities which involve more than one angle. Methods and examples of the use of the identities in simplifying expressions are given, and practice problems in simplification are included.

Also included in the chapter are methods for solving equations involving trigonometric functions. In many cases, the verification or simplification of an identity is an integral part of the solution of an equation. An additional topic considered in the chapter is the inverse trigonometric functions. Examples and problems involving equations and the inverse functions are also given.

Fundamental Identities

In earlier chapters it was shown that all of the trigonometric functions of an angle could be determined if one function or certain related

1

information was given. This seems to indicate that there are certain special relationships among the functions. These relationships are called identities and are independent of any particular angle. Many of the identities which will be considered in this section were established in earlier chapters and will be used here to change the form of an expression. In many problems, especially in calculus and other branches of mathematics, one particular method of expressing a function is more useful than any of the others. In these instances, the identities are used to put the expression in the desired form.

An equality which is true for all values of an unknown is called an identity. Identities are familiar in algebra, although they are not always specifically identified as such. A factoring process such as

$$(x^2 - 1) = (x - 1)(x + 1)$$

involves expression of an identity since it is true for all values of the variable. In trigonometric identities, the same situation must hold; that is, the equality must be true for all values of the variable.

Problems in identities are often given as equalities, and the identity is established by changing either one or both sides of the equality until both sides are the same. The fundamental rule in proving identities is as follows: NEVER WORK ACROSS THE EQUALITY SIGN. The algebraic rules for cross multiplication are never used. In this course all problems will be solved by working on only one side of the equality; that is, one side of the equality will be reduced or expanded until it is identical to the other side.

There are no hard and fast steps or methods to use in solving identities. However, there are some basic procedures or hints which will normally prove helpful in verification of the identities. Some of these hints are as follows:

1. Reduce the complex expressions to simple expressions, rather than building up from a simpler to a more complex one.
2. When possible, change the expression to one containing only sines and cosines.
3. If the expression contains fractions, it may help to change the form of the fractions.
4. Factoring an expression may suggest a subsequent step.
5. Keep the other member of the equality in mind. Since we are striving to change one expression to another, the form of the desired expression may suggest the steps to be taken.

Reciprocals and Ratios

One group of identities was presented in part in chapter 4 of this course. Recall that

$$\csc \theta = \frac{1}{\sin \theta}$$

Since

$$\sin \theta = \frac{y}{r}$$

and

$$\csc \theta = \frac{r}{y}$$

3

the cosecant function can be written as

$$\csc \theta = \frac{1}{\dfrac{y}{r}}$$

and simplified to

$$\csc \theta = \frac{1}{\sin \theta}$$

There are six reciprocal identities, one for each function. The numbers assigned these and subsequent identities in this chapter do not constitute any rules as to order or precedence. They are used only to simplify the explanation of steps in the example problems. The reciprocal formulas follow directly from the definitions of the trigonometric functions.

$$\sin \theta = \frac{1}{\csc \theta} \tag{1}$$

$$\cos \theta = \frac{1}{\sec \theta} \tag{2}$$

$$\tan \theta = \frac{1}{\cot \theta} \tag{3}$$

$$\csc \theta = \frac{1}{\sin \theta} \tag{4}$$

$$\sec \theta = \frac{1}{\cos \theta} \tag{5}$$

$$\cot \theta = \frac{1}{\tan \theta} \tag{6}$$

There are also identities involving the sine, cosine, tangent, and cotangent of an angle which are sometimes called ratio identities. These identities also result directly from the functions, and one of these expresses the tangent in terms of the sine and cosine, as follows:

$$\tan \theta = \frac{\sin \theta}{\cos \theta} \qquad (7)$$

Since
$$\sin \theta = \frac{y}{r}$$

and
$$\cos \theta = \frac{x}{r}$$

substituting these values in (7) gives

$$\tan \theta = \frac{\dfrac{y}{r}}{\dfrac{x}{r}}$$

which can be simplified to

$$\tan \theta = \frac{y}{x}$$

This is the definition of the tangent function given in an earlier chapter. The following identity for the cotangent,

$$\cot \theta = \frac{\cos \theta}{\sin \theta} \qquad (8)$$

can be shown to reduce identically to the definition of the cotangent function in terms of x, y, and r. The reciprocal and ratio identities are used to simplify trigonometric expressions as shown in the following example problems.

EXAMPLE: Simplify the expression

$$\sin \theta \, \cos \theta \, \tan \theta$$

to an expression containing only the sine function.

SOLUTION: One method of accomplishing this is to apply identity (7) to the given expression; then it becomes

$$\sin \theta \, \cos \theta \, \left(\frac{\sin \theta}{\cos \theta}\right)$$

or

$$\frac{\sin \theta \, \cos \theta \, \sin \theta}{\cos \theta}$$

Simplifying the cos θ terms in both the numerator and denominator of the fraction results in

$$\frac{\sin \theta \, \sin \theta}{1}$$

or $\sin^2 \theta$

This is the desired form and is identical, for all values of θ, to the original expression.

EXAMPLE: Use fundamental identities to verify the identity

$$\csc \theta + \cot \theta = \frac{1 + \cos \theta}{\sin \theta}$$

6

SOLUTION: Since the right-hand member of the identity contains sines and cosines, use (4) and (8) to change the left member to sines and cosines.
Then

$$\frac{1}{\sin \theta} + \frac{\cos \theta}{\sin \theta} = \frac{1 + \cos \theta}{\sin \theta}$$

Change the sum of fractions in the left member to a single fraction as in the following

$$\frac{1 + \cos \theta}{\sin \theta} = \frac{1 + \cos \theta}{\sin \theta}$$

and the identity is verified.

Observe that the right-hand member of the identity was not altered throughout the entire process. This is in accordance with our stated intention of working on just one side of the equality sign.

If we desire to verify this identity by retaining the left member and operating on the right member, the following steps may be used.

$$\csc \theta + \cot \theta = \frac{1 + \cos \theta}{\sin \theta}$$

Change the fraction in the right member to the sum of two fractions

$$\csc \theta + \cot \theta = \frac{1}{\sin \theta} + \frac{\cos \theta}{\sin \theta}$$

Next, apply (4) and (8) to the right member

$$\csc \theta + \cot \theta = \csc \theta + \cot \theta$$

and the identity is verified.

7

Squared Relationships

Another group of fundamental identities involves the squares of the functions. These, in some texts, are called Pythagorean identities since the Pythagorean theorem is used in their development. Consider the Pythagorean theorem

$$x^2 + y^2 = r^2$$

and divide both sides by r^2,

$$\frac{x^2}{r^2} + \frac{y^2}{r^2} = 1$$

Write this in the form

$$\left(\frac{x}{r}\right)^2 + \left(\frac{y}{r}\right)^2 = 1$$

and consider that

$$\cos \theta = \frac{x}{r}$$

and

$$\sin \theta = \frac{y}{r}$$

If $\cos \theta$ and $\sin \theta$ are substituted for $\frac{x}{r}$ and $\frac{y}{r}$ then

$$(\cos \theta)^2 + (\sin \theta)^2 = 1$$

This is rewritten as

8

$$\cos^2 \theta + \sin^2 \theta = 1$$

which is a fundamental squared or Pythagorean identity.

NOTE: The practice of writing an expression such as $(\sin \theta)^2$ in the form $\sin^2 \theta$ is common, and is the preferred method.

In the same manner, dividing both sides of the equation

$$x^2 + y^2 = r^2$$

by x^2 (where x is not equal to 0) gives

$$1 + \frac{y^2}{x^2} = \frac{r^2}{x^2}$$

or

$$1 + \left(\frac{y}{x}\right)^2 = \left(\frac{r}{x}\right)^2$$

Then, since $\tan \theta = \dfrac{y}{x}$

and $\sec \theta = \dfrac{r}{x}$

substitution gives $1 + (\tan \theta)^2 = (\sec \theta)^2$

or $$1 + \tan^2 \theta = \sec^2 \theta \qquad (10)$$

which is another fundamental identity.

The identity

$$1 + \cot^2 \theta = \csc^2 \theta \qquad (11)$$

9

is derived in a similar manner.

The three squared identities can be transposed algebraically to other forms with the following results:

$$\cos^2\theta = 1 - \sin^2\theta \qquad (12)$$

$$\sin^2\theta = 1 - \cos^2\theta \qquad (13)$$

$$\tan^2\theta = \sec^2\theta - 1 \qquad (14)$$

$$\sec^2\theta - \tan^2\theta = 1 \qquad (15)$$

$$\cot^2\theta = \csc^2\theta - 1 \qquad (16)$$

$$\csc^2\theta - \cot^2\theta = 1 \qquad (17)$$

In addition to the fundamental identities, there are many complicated identities involving the trigonometric functions. In the majority of cases, these identities can be proved by use of the laws of algebra and the fundamental identities.

EXAMPLE: Verify he identity

$$\frac{\sin\theta}{\csc\theta} + \frac{\cos\theta}{\sec\theta} = 1$$

SOLUTION: Reduce the left member to equal the right member. First, change each function to sines or cosines as follows:
Apply (4) to the denominator of the first fraction to obtain

$$\frac{\sin\,\theta}{\dfrac{1}{\sin\,\theta}} + \frac{\cos\,\theta}{\sec\,\theta} = 1$$

Simplification of the first fraction gives

$$\frac{\sin^2\theta}{1} + \frac{\cos\,\theta}{\sec\,\theta} = 1$$

Applying (5) to the remaining fraction gives

$$\sin^2\theta + \frac{\cos\,\theta}{\dfrac{1}{\cos\,\theta}} = 1$$

Simplification gives

$$\sin^2\theta + \cos^2\theta = 1$$

Then applying (9) to the left member results in

$$1 = 1$$

and the identity is verified.

EXAMPLE: Verify the following identity:

$$1 + \cot^2 2x = \frac{1}{\sin^2 2x}$$

SOLUTION: As a first step, apply (8) to the term $\cot^2 2x$.

Then
$$1 + \frac{\cos^2 2x}{\sin^2 2x} = \frac{1}{\sin^2 2x}$$

11

Combine the left term into a single fraction with a denominator of $\sin^2 2x$

$$\frac{\sin^2 2x + \cos^2 2x}{\sin^2 2x} = \frac{1}{\sin^2 2x}$$

Applying (9) to the numerator of the left member gives

$$\frac{1}{\sin^2 2x} = \frac{1}{\sin^2 2x}$$

and the identity is verified.

Practice Problems

Verify the following identities:

1. $\dfrac{1}{\tan^2 x + 1} = \cos^2 x$

2. $\csc x - \sin x = \cos x \cot x$

3. $\dfrac{\sin^2 \theta}{1 + \cos \theta} = 1 - \cos \theta$

4. $\cos \theta (\sec \theta - \cos \theta) = \sin^2 \theta$

5. $\tan^2 x (1 - \sin^2 x) = 1 - \cos^2 x$

6. $\sin^3 x = \dfrac{1 - \cos^2 x}{\csc x}$

7. $\dfrac{1}{2 + \cot^2 x} = \dfrac{1}{2\csc^2 x - \cot^2 x}$

12

Reduction Formulas

In chapter 4 of this course, reduction formulas were developed for dealing with angles greater than 90°. These reduction formulas can be combined into a general category of identities which also includes the formulas developed in chapter 4 for dealing with cofunctions and complementary angles. The formulas, of the type

$$\sin(90° - \theta) = \cos \theta$$

or $$\sec(180° + \theta) = -\sec \theta$$

are listed in chapter 4 and the listings will not be repeated in this chapter.

The formulas from chapter 4 will be used to simplify expressions, in the same manner as the other identities, in the following examples.

EXAMPLE: Simplify the expression

$$\sin(180° - \theta)\tan(90° - \theta)\cot(180° - \theta)$$

into an expression containing functions of θ alone.

SOLUTION: From chapter 4, the following formulas are chosen

$$\sin(180° - \theta) = \sin \theta$$

$$\tan(90° - \theta) = \cot \theta$$

$$\cot(180° - \theta) = -\cot \theta$$

Substitution of these values in the expression

$$\sin(180° - \theta)\tan(90° - \theta)\cot(180° - \theta)$$

13

results in the expression

$$\sin \theta \, \cot \theta \, (-\cot \theta)$$

Rewrite this in the form

$$-\sin \theta \, \cot \theta \, \cot \theta$$

and apply identity (8) to one of the cot θ factors.

Then, $\qquad -\sin \theta \left(\dfrac{\cos \theta}{\sin \theta}\right) \cot \theta$

results and this can be simplified to

$$-\cos \theta \, \cot \theta$$

to complete the problem.

EXAMPLE: Express the following as an expression containing the least possible number of functions of θ.

$$\sin(360° - \theta)\tan(90° - \theta)\csc \theta$$

SOLUTION: The following formulas are given in chapter 4:

$$\sin(360° - \theta) = \sin(-\theta) = -\sin \theta$$

and $\quad \tan(90° - \theta) = \cot \theta$

Substitution of these values in the original expression results in

$$-\sin \theta \, \cot \theta \, \csc \theta$$

Rewrite this as

$$-\cot\theta\ \sin\theta\ \csc\theta$$

and apply identity (1) to the factor $\sin\theta$ to arrive at

$$-\cot\theta\ \csc\theta\ \frac{1}{\csc\theta}$$

or $\qquad\qquad -\cot\theta$

which is an expression in terms of one function of θ.

Practice Problems

Express the following as expressions containing the least number of functions as θ.

1. $\sin(180° - \theta)\ \sin\theta$

2. $\sin(180° + \theta)\ \sec(180° - \theta)$

3. $\cos(360° - \theta)\ \cot(90° - \theta)\ \csc(90° - \theta)$

Answers

1. $\sin^2\theta$
2. $\tan\theta$
3. $\tan\theta$

Inverse Trigonometric Functions

In this section we will discuss the definitions which apply to the inverse trigonometric functions along with the principal values of these functions. Relations among these functions will be examined by the use of examples and practice problems.

Definitions

It is often convenient and useful to turn a trigonometric function around so that instead of writing

$$\tan \theta = A$$

we write

$$\theta = \text{the angle whose tangent is A}$$

Rather than write out the last statement, mathematicians use either of the following notations:

$$\theta = \arctan A$$

or
$$\theta = \tan^{-1} A$$

In the last notation we do not mean -1 to represent an algebraic exponent and $\tan^{-1}A$ does not denote $\dfrac{1}{\tan A}$. If we meant $\tan^{-1}A$ equals $\dfrac{1}{\tan A}$, we would have written $(\tan A)^{-1}$ equals $\dfrac{1}{\tan A}$.

In this course, the preferred notation is arctan A.

Principal Values

For any angle there is one and only one function which corresponds to it; but to any value of a trigonometric function, there are numerous angles which will satisfy the value. For instance,

$$\theta = \arctan 1$$

16

can be written

$$\tan \theta = 1$$

but 1 is the tangent of many angles such as 45°, 225°, 405°, 585°, and others. Any angle θ which satisfies (45° + n . 180°), where n is an integer, satisfies the expression

$$\tan \theta = 1$$

For any inverse trigonometric function there are two angles less than 360° which satisfy it. Thus,

θ = arccos(0.500) refers to 60° and 300°

θ = arccos(-0.500) refers to 120° and 240°

θ = arcsin(0.707) refers to 45° and 135°

θ = arcsec(2.000) refers to 60° and 300°

Since a given inverse trigonometric function has many values, one of these values is selected as its principal value.

To denote principal values, we will capitalize the first letter in the name and we will use the ranges for principal values as follows:

$$-90° \leq \text{Arcsin } x \leq 90°$$

$$0° \leq \text{Arccos } x \leq 180°$$

$$-90° < \text{Arctan } x < 90°$$

$$-90° \leq \text{Arccsc } x \leq 90°$$

$$0° \leq \text{Arcsec } x \leq 180°$$

$$0° < \text{Arccot } x < 180°$$

All principal values lie between -90° and 180° proceeding counterclockwise from -90°.

The principal values for positive numbers are between 0° and 90°. For negative numbers, principal values of the inverse sine, tangent, and cosecant lie between -90° and 0°, while principal values of the inverse cosine, cotangent, and secant lie between 90° and 180°. We will use, for understanding, the examples which follow.

EXAMPLE: Find the principal value of the angle in the function

$$\theta = \text{Arccos}\ (0.4472)$$

SOLUTION: Using the trigonometric tables, we find the angle whose cosine is 0.4472 is 63° 26' or 296° 34'. We choose 63° 26' as this is the first quadrant angle and is the principal value. We reject 296° 34' because it is a fourth quadrant angle and the principal values for the Arccos function are limited to the first and second quadrants.

EXAMPLE: Find the principal value of the angle in the function

$$\theta = \text{Arccos}\ (-0.5000)$$

SOLUTION: Using the trigonometric tables, we find the angle whose cosine is (-0.5000) is 120° or 240°. We choose 120° because 240° is in the third quadrant and does not satisfy the value we agreed on as the principal value for the cosine function.

EXAMPLE: Find the principal value of the angle in the function

$$\theta = \text{Arctan}\ 1$$

SOLUTION: The angle whose tangent is 1 is 45° or 225°. We reject 225°, a third quadrant angle, and select 45° because it is a first quadrant angle.

EXAMPLE: Find the principal value of the angle in the function

$$\theta = \text{Arcsec } 2.236$$

SOLUTION: If the trigonometric tables do not list secant values, the function may be changed by the following:

Since $\qquad \sec \theta = \dfrac{1}{\cos \theta}$

then $\quad$ Arcsec $(2.236) = $ Arccos $\dfrac{1}{2.236}$

$$= \text{Arccos } (0.4472)$$

and we find that the angle whose cosine is 0.4472 is 63° 26' and the principal value of the angle whose secant is 2.236 is 63° 26'.

Practice Problems

Find the principal values of the angles in the following functions:

1. $\theta = $ Arccos (0.9135)

2. $\theta = $ Arcsin (0.8829)

3. $\theta = $ Arctan (11.430)

4. $\theta = $ Arccot (-0.1169)

5. $\theta = $ Arcsec (1.0075)

6. $\theta = $ Arctan (-0.1228)

19

Answers

1. $24°$
2. $62°$
3. $85°$
4. $96° \ 40'$
5. $7°$
6. $-7°$

In dealing with the inverse trigonometric functions, we may be presented with the problem of finding the principal value of the function $\theta = \text{Arcsin}\left(\dfrac{\sqrt{2}}{2}\right)$. In this case we could solve it as follows:

Draw a right triangle as shown in figure 8-1. This expression $\dfrac{\sqrt{2}}{2}$ can be rewritten as $\dfrac{1}{\sqrt{2}}$ by the following steps:

$$\frac{\sqrt{2}}{2}\left(\frac{\sqrt{2}}{\sqrt{2}}\right) = \frac{2}{2\sqrt{2}} = \frac{1}{\sqrt{2}}$$

Recall that $\sin \theta$ equals $\dfrac{y}{r}$. The triangle is a 45°-90° triangle as shown previously. Now, the function

$$\theta = \text{Arcsin} \ \frac{\sqrt{2}}{2}$$

20

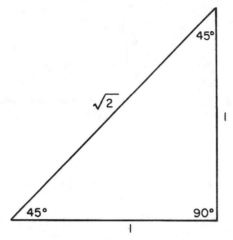

Figure 8-1. – 45° - 90° triangle.

becomes $\qquad \theta = \text{Arcsin } \dfrac{1}{\sqrt{2}}$

and we find that $\qquad \theta = 45°$

Use this same approach to answer the following questions.

Practice Problems

Find the principal values of the angles in the following functions.

1. $\theta = \text{Arccos}\left(\dfrac{\sqrt{2}}{2}\right)$

2. $\theta = \text{Arctan}\,(\sqrt{3})$

3. $\theta = \text{Arccot}\,(-\sqrt{3})$

4. $\theta = \text{Arccos}\left(\dfrac{\sqrt{3}}{2}\right)$

21

5. $\theta = \text{Arcsin} \left(\dfrac{\sqrt{3}}{2} \right)$

6. $\theta = \text{Arccot} \left(-\dfrac{1}{\sqrt{3}} \right)$

Answers

1. $45°$

2. $60°$

3. $150°$

4. $30°$

5. $60°$

6. $120°$

If we are to find, using the principal values, the value of the expression Arctan $\sqrt{3}$ - Arctan $\dfrac{1}{\sqrt{3}}$ in radians, we proceed as follows:

$$\text{Arctan } \sqrt{3} = 60°$$

and

$$\text{Arctan } \dfrac{1}{\sqrt{3}} = 30°$$

thus

$$\text{Arctan } \sqrt{3} - \text{Arctan } \dfrac{1}{\sqrt{3}} = 60° - 30°$$

and

$$1° = \dfrac{\pi}{180} \text{ radians}$$

then

$$30° = \dfrac{\pi}{180} \left(\dfrac{30}{1} \right) \text{ radians}$$

$$= \dfrac{\pi}{6} \text{ radians}$$

Practice Problems

Using the principal values, give the values of the following expressions in radians:

1. $\text{Arcsin } \dfrac{1}{2} - \text{Arccos } \dfrac{1}{2}$

2. $\text{Arccos } \dfrac{\sqrt{3}}{2} - \text{Arcsin } \dfrac{\sqrt{3}}{2}$

3. $\text{Arctan } 1 - \text{Arctan } \dfrac{1}{\sqrt{3}}$

4. $\text{Arctan } \sqrt{3} - \text{Arcsin } \dfrac{1}{2}$

Answers

1. $-\dfrac{\pi}{6}$

2. $-\dfrac{\pi}{6}$

3. $\dfrac{\pi}{12}$

4. $\dfrac{\pi}{6}$

Relations Among Inverse Functions

In order to understand the relations among the inverse functions, we will start by drawing a triangle. If θ equals Arcsin x, we can write sin θ equals x. We now draw a triangle which contains the angle whose sine is x and assume the hypotenuse equal to one. The remaining side of the triangle will be, from the Pythagorean theorem, $\sqrt{1 - x^2}$. This is shown in figure 8-2.

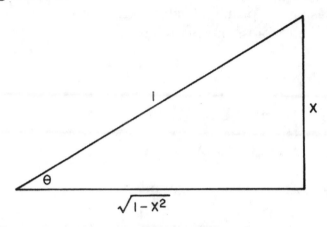

Figure 8-2. – Triangle containing Arcsin x.

Now, we can write all of the functions and inverse functions of the angle θ in terms of the sides of the triangle as follows:

$$\sin \theta = x, \qquad \text{or } \theta = \text{Arcsin } x$$

$$\cos \theta = \sqrt{1 - x^2}, \text{ or } \theta = \text{Arccos } \sqrt{1 - x^2}$$

$$\tan \theta = \frac{x}{\sqrt{1 - x^2}}, \text{ or } \theta = \text{Arctan } \frac{x}{\sqrt{1 - x^2}}$$

$$\csc \theta = \frac{1}{x}, \qquad \text{or } \theta = \text{Arccsc } \frac{1}{x}$$

$$\sec \theta = \frac{1}{\sqrt{1 - x^2}}, \text{ or } \theta = \text{Arcsec } \frac{1}{\sqrt{1 - x^2}}$$

$$\cot \theta = \frac{\sqrt{1 - x^2}}{x}, \text{ or } \theta = \text{Arccot} \frac{\sqrt{1 - x^2}}{x}$$

All of the inverse functions are equal to θ; there-fore, they are equal to each other. We will use this type of analysis to solve a few problems.

EXAMPLE: Using principal values, find the tangent of the angle whose sine is $\frac{\sqrt{3}}{2}$; that is,

$$\tan \text{Arcsin} \frac{\sqrt{3}}{2} = ?$$

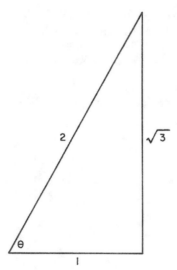

Figure 8-3. – Triangle containing Arcsin $\frac{\sqrt{3}}{2}$.

SOLUTION: Draw the triangle containing the Arcsin $\frac{\sqrt{3}}{2}$ as shown in figure 8-3. The remaining side will be given by

25

$$x = \sqrt{2^2 - (\sqrt{3})^2} = \sqrt{1} = 1$$

Using the tangent ratio we have

$$\tan \theta = \sqrt{3}$$

and

$$\theta = 60°$$

EXAMPLE: Using principal values find

$$\sin\left(\text{Arccos } \frac{3}{5} - \text{Arcsin } \frac{4}{5}\right)$$

SOLUTION: Draw the triangle as shown in figure 8-4. The missing side of the triangle containing Arccos $\frac{3}{5}$ is given by

$$y = \sqrt{5^2 - 3^2} = \sqrt{16} = 4$$

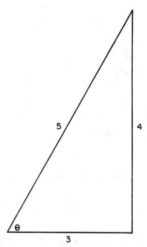

Figure 8-4. – Triangle containing Arccos $\frac{3}{5}$.

Notice that this triangle also contains $\text{Arcsin}\frac{4}{5}$, so that

$$\text{Arccos}\,\frac{3}{5} = \text{Arcsin}\,\frac{4}{5}$$

and the original expression becomes

$$\sin 0°$$

which is zero.

Practice Problems

Evaluate the following expressions:

1. $\sin\left(\text{Arccos}\,\frac{3}{5}\right)$

2. $\tan\left(\text{Arcsin}\,\frac{1}{10}\right)$

3. $\cot\left(\text{Arcsin}\,\frac{x}{1+x}\right)$

4. $\sec\left(\text{Arctan}\,\frac{x^2}{x^2-1}\right)$

5. $\cos\left(\text{Arcsin}\,\frac{1}{x^2-1}\right)$

Answers

1. $\frac{4}{5}$

2. $\frac{1}{9.95}$

3. $\dfrac{\sqrt{2x + 1}}{x}$

4. $\dfrac{\sqrt{2x^4 - 2x^2 + 1}}{x^2 - 11}$

5. $\dfrac{x\sqrt{x^2 - 2}}{x^2 - 1}$

Formulas

In this section we will discuss the trigonometric formulas for addition and subtraction of angles, half angles, double angles, and transcendental functions. We will use examples for better understanding and in some instances we will derive formulas.

Addition and Subtraction Formulas

Four formulas express the sine or cosine of the sum and difference of two angles as a function of the sines and cosines of the single angles. They are very important because they are the basis for much of trigonometric analysis. From the following four formulas we may derive all of the formulas in the following sections:

$$\sin (A + B) = \sin A \cos B + \cos A \sin B \quad (1)$$

$$\cos (A + B) = \cos A \cos B - \sin A \sin B \quad (2)$$

$$\sin (A - B) = \sin A \cos B - \cos A \sin B \quad (3)$$

$$\cos (A - B) = \cos A \cos B + \sin A \sin B \quad (4)$$

We will prove these four formulas for angles whose sum is less than 90°. They are actually true for all angles.

In figure 8-5 we have indicated the sum of two angles, A and B. The hypotenuse of the triangle containing angle B has been set equal to 1 so that the legs of this triangle have the values sin B and cos B.

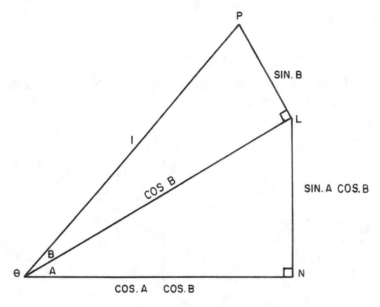

Figure 8-5. – Sum of two angles, part one.

Thus, $\dfrac{PL}{1} = \sin B$

and $\dfrac{OL}{1} = \cos B$

In the triangle containing angle A, we can see that

$$\cos A = \frac{ON}{OL} = \frac{ON}{\cos B}$$

Therefore, $ON = \cos A \cos B$ (5)

and $\sin A = \frac{NL}{OL} = \frac{NL}{\cos B}$

Therefore $NL = \sin A \cos B$ (6)

Now, let us add a construction to the figure, as in figure 8-6, and calculate more lines in

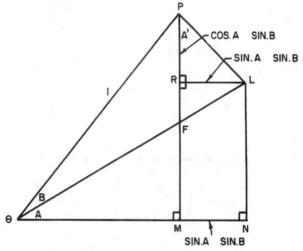

Figure 8-6. – Sum of two angles, part two.

terms of the sine and cosine of angles A and B. First, let us prove that angle A is equal to angle A'.

Triangle OMF is similar to triangle PFL, and angles A and A' are corresponding angles. Therefore, angle A equals angle A'. In triangle PRL

$$\cos A' = \frac{PR}{\sin B}$$

and $$PR = \cos A' \sin B$$

$$= \cos A \, \sin B \qquad (7)$$

Also, $$\sin A' = \frac{RL}{\sin B}$$

and $$RL = \sin A' \sin B$$

$$= \sin A \, \sin B \qquad (8)$$

Furthermore,

$$RL = MN = \sin A \sin B \qquad (9)$$

Now, in triangle OMP we can write

$$\sin (A + B) = \frac{PM}{1} = PR + RM \qquad (10)$$

but $$RM = LN \qquad (11)$$

so $$\sin(A + B) = PR + LN = LN + PR \qquad (12)$$

Also $$\cos(A + B) = \frac{OM}{1} = ON - MN \qquad (13)$$

Substituting from equations (6) and (7) into equation (12)

$$\sin(A + B) = \sin A \cos B + \cos A \sin B \qquad (14)$$

31

Substituting from equations (5) and (9) into equation (13)

$$\cos(A + B) = \cos A \cos B - \sin A \sin B \quad (15)$$

EXAMPLE: Use the addition formulas to find cos 75°.

SOLUTION: Use cos(A + B) equal to cos A cos B - sin A sin B. From this we write cos 75° equals cos (45° + 30°) and substitute as follows:

$$\cos(45° + 30°) = \cos 45° \cos 30° - \sin 45° \sin 30°$$

$$= \left(\frac{1}{\sqrt{2}}\right)\left(\frac{\sqrt{3}}{2}\right) - \left(\frac{1}{\sqrt{2}}\right)\left(\frac{1}{2}\right)$$

$$= \left(\frac{\sqrt{2}}{2}\right)\left(\frac{\sqrt{3}}{2}\right) - \left(\frac{\sqrt{2}}{2}\right)\left(\frac{1}{2}\right)$$

$$= \frac{\sqrt{6} - \sqrt{2}}{4}$$

The subtraction formulas can be derived from the addition formulas by substituting (-B) for (+B). Now, we have

$$\sin(A - B) = \sin A \cos(-B) + \cos A \sin(-B) \quad (16)$$

and

$$\cos(A - B) = \cos A \cos(-B) - \sin A \sin (-B) \quad (17)$$

But, the cosine of a negative angle is equal to the cosine of the angle. The sine of a negative angle, however, is equal to minus the sine of the angle; that is,

$$\cos(-B) = \cos B \qquad (18)$$

$$\sin(-B) = -\sin B \qquad (19)$$

Substitution of these values in equations (16) and (17) gives

$$\sin(A - B) = \sin A \cos B - \cos A \sin B \qquad (20)$$

and

$$\cos(A - B) = \cos A \cos B + \sin A \sin B \qquad (21)$$

EXAMPLE: Show that

$$\sin(45° - \theta) = \frac{\cos \theta - \sin \theta}{\sqrt{2}}$$

SOLUTION: Applying equation (20)

$$\sin(45° - \theta) = \sin 45° \cos \theta - \cos 45° \sin \theta$$

but

$$\sin 45° = \cos 45° = \frac{1}{\sqrt{2}}$$

Substituting these values we have

$$\sin(45° - \theta) = \frac{\cos \theta}{\sqrt{2}} - \frac{\sin \theta}{\sqrt{2}}$$

$$= \frac{\cos \theta - \sin \theta}{\sqrt{2}}$$

We can use these four formulas, (14), (15), (20), and (21), to drive a number of other important ones. One of these is the tangent addition formula.

33

$$\tan(A + B) = \frac{\tan A + \tan B}{1 - \tan A \tan B} \qquad (22)$$

In order to prove this formula, proceed as follows:

Taking the ratio of equalities (14) and (15), we have

$$\frac{\sin(A + B)}{\cos(A + B)} = \frac{\sin A \cos B + \cos A \sin B}{\cos A \cos B - \sin A \sin B} \qquad (23)$$

Dividing both numerator and denominator of the right side of this equation by cos A cos B we have

$$\tan(A + B) = \frac{\dfrac{\sin A}{\cos A} + \dfrac{\sin B}{\cos B}}{1 - \dfrac{\sin A \sin B}{\cos A \cos B}} \qquad (24)$$

which reduces to

$$\tan(A + B) = \frac{\tan A + \tan B}{1 - \tan A \tan B} \qquad (25)$$

Another important formula is the subtraction formula for the tangent. To find tan(A - B) replace B with (-B). Therefore,

$$\tan(A - B) = \frac{\tan A + \tan(-B)}{1 - \tan A \tan(-B)} \qquad (26)$$

$$= \frac{\tan A - \tan B}{1 + \tan A \tan B} \qquad (27)$$

EXAMPLE: Use subtraction formulas to find tan 15°

SOLUTION: Use the special triangles as previously discussed.

$$\tan 15° = \tan(45° - 30°)$$

and

$$\tan(45° - 30°) = \frac{\tan 45° - \tan 30°}{1 + \tan 45° \tan 30°}$$

$$= \frac{1 - \frac{\sqrt{3}}{3}}{1 + (1)\frac{\sqrt{3}}{3}}$$

$$= \frac{3 - \sqrt{3}}{3 + \sqrt{3}}$$

$$= \frac{3 - \sqrt{3}}{3 + \sqrt{3}}\left(\frac{3 - \sqrt{3}}{3 - \sqrt{3}}\right)$$

$$= \frac{9 - 6\sqrt{3} + 3}{9 - 3}$$

$$= \frac{12 - 6\sqrt{3}}{6}$$

$$= 2 - \sqrt{3}$$

EXAMPLE: Given $\tan 45°$ equals 1 and $\tan 60°$ equals $\sqrt{3}$, find the tangent of 105°.

SOLUTION: Applying this knowledge to equation (22), we have

$$\tan(45° + 60°) = \tan 105° = \frac{1 + \sqrt{3}}{1 + \sqrt{3}}$$

It is easy to evaluate a fraction in this form by rationalizing the denominator.

Multiply the numerator and denominator by the same numbers as they appear in the denominator but connected by the by the apposite sign.

$$\frac{(1 + \sqrt{3})^2}{(1 - \sqrt{3})(1 + \sqrt{3})} = \frac{1 + 2\sqrt{3} + 3}{1 - 3}$$

$$= \frac{4 + 2\sqrt{3}}{-2}$$

$$= -(2 + \sqrt{3})$$

$$= -3.732$$

Therefore, $\tan 105° = -3.732$

Practice Problems

Use the addition and subtraction formulas to find the values of the following without tables:

1. $\sin 75°$

2. $\cos 15°$

3. $\tan 75°$

4. $\cot 165°$ Hint: recall $\cot(180 - \theta) = -\cot \theta$ and $\cot \theta = \dfrac{1}{\tan \theta}$

Answers

1. $\dfrac{\sqrt{6} + \sqrt{2}}{4}$ 3. $2 + \sqrt{3}$

2. $\dfrac{\sqrt{6} + \sqrt{2}}{4}$ 4. $\dfrac{1}{\sqrt{3} - 2}$ or $-\sqrt{3} - 2$

Double Angle Formulas

The addition formulas may be used to derive the double angle formulas.

$$\sin 2A = 2 \sin A \cos A$$

$$\cos 2A = \cos^2 A - \sin^2 A$$

$$\tan 2A = \frac{2 \tan A}{1 - \tan^2 A} \qquad (28)$$

In equations (14) and (15), if B equals A, we can write

$$\sin 2A = \sin A \cos A + \cos A \sin A \qquad (29)$$
$$\cos 2A = \cos A \cos A - \sin A \sin A$$

from which we obtain

$$\sin 2A = 2 \sin A \cos A \qquad (30)$$

and, using $\sin^2 A + \cos^2 A = 1$

then $\cos 2A = \cos^2 A - \sin^2 A$

$$= 2 \cos^2 A - 1 \qquad (31)$$

$$= 1 - 2 \sin^2 A$$

37

Substituting A for B in equation (25), we obtain

$$\tan 2A = \frac{2 \tan A}{1 - \tan^2 A} \qquad (32)$$

EXAMPLE: Evaluate $\sin 15° \cos 15°$.

SOLUTION: Since

$$2 \sin A \cos A = \sin 2A$$

and

$$\sin A \cos A = \frac{1}{2} \sin 2A$$

then

$$\sin 15° \cos 15° = \frac{1}{2} \sin 2(15°)$$

$$= \frac{1}{2} \sin 30°$$

$$= \left(\frac{1}{2}\right)\left(\frac{1}{2}\right)$$

$$= \frac{1}{4}$$

EXAMPLE: Find the three first quadrant angles which satisfy the trigonometric equation

$$\sin 4x = \cos 2x$$

SOLUTION: From the double angle formulas, we can write

$$2 \sin 2x \cos 2x = \cos 2x$$

or

$$\cos 2x(2 \sin 2x - 1) = 0$$

The solutions of this equation may be obtained by setting the factors equal to zero and making use of inverse trigonometric functions. We may write

$$\cos 2x = 0$$
$$2x = \text{Arccos } 0$$
$$= 90°$$
$$x = 45°$$

and

$$2 \sin 2x - 1 = 0$$
$$\sin 2x = \frac{1}{2}$$
$$2x = \text{Arcsin } \frac{1}{2}$$
$$2x = 30°, \ 150°$$
$$x = 15°, \ 75°$$

The equation has three solutions, x equals 15°, 45°, and 75°. Notice that in writing down the values of the inverse functions, it was necessary to include 150° since, when divided by 2, this gives an angle in the first quadrant.

Half Angle Formulas

Dividing all the angles in equation (31) by 2 we obtain

$$\cos A = \cos^2 \frac{A}{2} - \sin^2 \frac{A}{2} \tag{33}$$

Using equation (33), we can derive two useful

39

formulas. Adding and subtracting $\sin^2 \frac{A}{2}$ on the right side of equation (33) we have

$$\cos A = (\cos^2 \frac{A}{2} + \sin^2 \frac{A}{2}) - 2 \sin^2 \frac{A}{2} \qquad (34)$$

Observe that the methods necessary for simplifying trigonometric identities and equations often include operations which may at first appear to be pointless. In the preceding sentence we referred to "adding and subtracting $\sin^2 \frac{A}{2}$ on the right side of equation (33)." The advantage of adding $\sin^2 \frac{A}{2}$ becomes obvious when we group it with $\cos^2 \frac{A}{2}$. The expression $(-\sin^2 \frac{A}{2})$ is added to the right-hand side along with $(\sin^2 \frac{A}{2})$ in order to avoid changing the overall value.

The quantity in the parentheses in equation (34) is equal to 1, so

$$\cos A = 1 - 2 \sin^2 \frac{A}{2} \qquad (35)$$

Rearranging equation (35) and taking the square root of both sides, we have

$$\sin \frac{A}{2} = \pm \sqrt{\frac{1 - \cos A}{2}} \qquad (36)$$

Adding and subtracting $\cos^2 \frac{A}{2}$ on the right side of equation (33), we have

$$\cos A = 2 \cos^2 \frac{A}{2} - \left(\cos^2 \frac{A}{2} + \sin^2 \frac{A}{2}\right) \qquad (37)$$

but, the quantity within the parentheses is equal to 1 so that

$$\cos A = 2 \cos^2 \frac{A}{2} - 1 \qquad (38)$$

Rearranging equation (38) and taking the square root of both sides gives us

$$\cos \frac{A}{2} = \pm \sqrt{\frac{1 + \cos A}{2}} \qquad (39)$$

Taking the ratio of equations (36) and (39), we have

$$\tan \frac{A}{2} = \pm \sqrt{\frac{1 - \cos A}{1 + \cos A}} \qquad (40)$$

Notice that the use of (+) or (-) is dependent upon the quadrant of angle termination.

EXAMPLE: Find cos 15°, if cos 30° equals 0.866 or $\frac{\sqrt{3}}{2}$.

SOLUTION: From equation (39) we have

$$\cos \frac{30°}{2} = \cos 15° = \sqrt{\frac{1 + 0.866}{2}} = \sqrt{0.933}$$

Thus $\qquad \cos 15° = 0.9659$

The solution using $\frac{\sqrt{3}}{2}$ follows:

$$\cos \frac{30°}{2} = \cos 15° = \sqrt{\frac{1 + \frac{\sqrt{3}}{2}}{2}}$$

$$= \sqrt{\frac{\frac{2 + \sqrt{3}}{2}}{2}}$$

$$= \frac{\sqrt{2} + \sqrt{3}}{2}$$

41

Practice Problems

Use the half angle formulas to find the exact value of the following:

1. $\sin 15°$

2. $\cos 135°$

3. $\tan 22.5°$

4. $\tan 195°$

Answers

1. $\pm \dfrac{\sqrt{2 - \sqrt{3}}}{2}$

2. $\pm \sqrt{\dfrac{1}{2}}$

3. $\pm \sqrt{3 - 2\sqrt{2}}$

4. $\pm \sqrt{7 - 4\sqrt{3}}$

Transcendental Functions

To define transcendental functions we state that any functions other than algebraic functions will be classified, for the purposes of this course, as transcendental functions. This group of transcendental functions includes such functions as trigonometric functions, inverse trigonometric functions, exponential functions, and logarithmic functions.

Later in calculus we will prove that the sine and cosine of an angle can be calculated from the following series, if the angle is expressed in radian measure:

$$\sin x = x - \frac{x^3}{3!} + \frac{x^5}{5!} - \frac{x^7}{7!} + \cdots$$

and

$$\cos x = 1 - \frac{x^2}{2!} + \frac{x^4}{4!} - \frac{x^6}{6!} + \cdots$$

NOTE: 3! is read 3 factorial and is equal to 1 x 2 x 3 or 6. 5! is read 5 factorial and is equal to 1 x 2 x 3 x 4 x 5 or 120.

At the same time the expansion of the function e^x where e is the number 2.71828. . . , the base of the system of natural logarithms, is as follows:

$$e^x = 1 + x + \frac{x^2}{2!} + \frac{x^3}{3!} + \frac{x^4}{4!} + \cdots$$

and

$$e^{-x} = 1 - x + \frac{x^2}{2!} - \frac{x^3}{3!} + \frac{x^4}{4!} - \cdots$$

Moreover, by using the notation $i = \sqrt{-1}$ two similar expressions are obtained:

$$e^{ix} = 1 + ix - \frac{x^2}{2!} - \frac{ix^3}{3!} + \frac{x^4}{4!} + \frac{ix^5}{5!} - \frac{x^6}{6!} - \frac{ix^7}{7!} + \cdots$$

and

$$e^{-ix} = 1 - ix - \frac{x^2}{2!} + \frac{ix^3}{3!} + \frac{x^4}{4!} - \frac{ix^5}{5!} - \frac{x^6}{6!} + \frac{ix^7}{7!} + \cdots$$

43

Adding e^{ix} to e^{-ix} term by term and dividing by two we obtain an expression equal to cos x.

$$\frac{e^{ix} + e^{-ix}}{2} = \cos x$$

Subtracting e^{-ix} from e^{ix} term by term and dividing by 2i we have an expression equal to sin x.

$$\frac{e^{ix} - e^{-ix}}{2i} = \sin x$$

In calculus, trigonometric and logarithmic functions are grouped together and called transcendental functions, partly because they cannot be expressed by a simple algebraic formula, and partly because they both are related to the number e. This relation becomes important in derivations in advanced and applied calculus.

Equations

A trigonometric equation is an equality which is true for some values but may not be true for all values of the variable. The principles and processes used to solve algebraic equations may be used to solve trigonometric equations. The identities and reduction formulas previously studied may also be used in solving trigonometric equations. There are so many different approaches to solving these equations that we will use several examples for better understanding.

Multiple Solutions

We will use the following examples and practice problems to show the multiple solutions of a trigonometric equation.

EXAMPLE: Find the value of θ if

$$\tan \theta = 1$$

SOLUTION: We must find the angle or angles having a tangent equal to 1. Using the inverse trigonometric functions, we may write

$$\tan \theta = 1$$

$$\theta = \arctan 1$$

$$\theta = 45°$$

and recalling that

$$\tan(180° + \theta) = \tan \theta$$

then
$$\tan(180° + 45°) = \tan 45°$$

$$\tan 225° = \tan 45°$$

Thus,
$$\arctan 1 = 45°$$

and
$$\arctan 1 = 225°$$

Therefore, the solutions for the equation are 45° and 225°. The 45° angle is a first quadrant angle and the 225° angle is a third quadrant angle and both have a positive sign and the same trigonometric value; thus we have a multiple solution.

45

Notice that the two solutions of the equation differ from the investigation of the inverse trigonometric functions in which we were required to find the PRINCIPAL value. The PRINCIPAL value solution was found to be in a particular quadrant. In multiple solutions we will use the term PRIMARY to indicate that we are searching for solutions which are restricted to the range

$$0° \leq \theta < 360°$$

Notice also that if we remove the restriction of the term PRIMARY we may write

$$\tan \theta = \tan(\theta + n \cdot 360°)$$

and we find, if n is an integer, that there are many solutions to the equation.

EXAMPLE: Find the primary solutions to the equation

$$\cos \theta = \frac{\sqrt{3}}{2}$$

SOLUTION: We first use the inverse trigonometric function to write

$$\cos \theta = \frac{\sqrt{3}}{2}$$

then
$$\theta = \arccos \frac{\sqrt{3}}{2}$$
$$= 30°$$

but
$$\cos \theta = \cos (360° - \theta)$$

$$\cos 30° = \cos 330°$$

Therefore, the solutions are 30° and 330°.

EXAMPLE: Find the primary solutions to the equation

$$\csc \theta = 2$$

SOLUTION: As in the previous example, we write

$$\csc \theta = 2$$

and

$$\csc \theta = \frac{1}{\sin \theta}$$

$$\frac{1}{\sin \theta} = 2$$

$$\sin \theta = \frac{1}{2}$$

then

$$\theta = \arcsin \frac{1}{2}$$

$$= 30°$$

but

$$\sin \theta = \sin (180° - \theta)$$

$$\sin 30° = \sin 150°$$

Therefore, the solutions are 30° and 150°

Practice Problems

Find the primary values of θ in the following equations:

1. $\tan \theta = -1$

2. $\sec \theta = 2$

3. $\sin \theta = -\dfrac{\sqrt{3}}{2}$

4. $\cos \theta = \dfrac{1}{2}$

Answers

1. 135° and 315°

2. 60° and 300°

3. 240° and 300°

4. 60° and 300°

The following examples show how to find the solution of more difficult equations.

EXAMPLE: Find the primary values of the equation

$$\sin \theta + \sin \theta \cot \theta = 0$$

SOLUTION: We factor the equation and find that

$$\sin \theta + \sin \theta \cot \theta = 0$$

and $\qquad \sin \theta \,(1 + \cot \theta) = 0$

Setting each factor equal to zero, we have the two equations

$$\sin \theta = 0$$

48

and $\qquad$ $1 + \cot \theta = 0$

Solve each as follows:

$$\sin \theta = 0$$

then $\qquad$ $\theta = \arcsin 0$

$$= 0° \text{ and } 180°$$

and $\qquad$ $1 + \cot \theta = 0$

$$\cot \theta = -1$$

Therefore,

$$\theta = \text{arccot} -1$$

$$= 135° \text{ and } 315°$$

Therefore, the solutions to the equation are 0°, 135°, 180°, and 315°.

EXAMPLE: Find the primary values of the equation

$$2 \sin^2 \theta + \sin \theta - 3 = 0$$

SOLUTION: Factor the equation and then set each factor equal to zero, as follows:

$$2 \sin^2 \theta + \sin \theta - 3 = 0$$

and $\qquad$ $(2 \sin \theta - 1)(\sin \theta - 1) = 0$

then $\qquad$ $2 \sin \theta - 1 = 0$

and $\qquad$ $\sin \theta - 1 = 0$

Solve each as follows:

$$2 \sin \theta - 1 = 0$$

$$2 \sin \theta = 1$$

$$\sin \theta = \frac{1}{2}$$

then $\qquad \theta = \arcsin \frac{1}{2}$

and $\qquad \theta = 30° \text{ and } 150°$

Also, $\qquad \sin \theta - 1 = 0$

$$\sin \theta = 1$$

then $\qquad \theta = \arcsin 1$

$$= 90°$$

Therefore, the solutions are **30°, 90°, and 150°.**

Practice Problems

Find the primary value of the following equations:

1. $2 \cos^2 \theta = 3 \cos \theta - 1$

2. $\tan^2 \theta = 3$

3. $2 \sin^2 \theta - 3 \sin \theta = -1$

Answers

1. 0°, 60°, 300°

2. 60°, 120°, 240°, 300°

3. 30°, 90°, 150°

Limited Solutions

We will consider two types of possible solutions to fall within the category of limited solutions. The following examples show both of these types.

The first type of limited solution occurs when an equation is solved and one of the solutions is not true upon inspection.

EXAMPLE: Find the primary values of θ in the equation

$$\sin^2\theta \, \sec \theta = \sec \theta$$

SOLUTION: We first rearrange, and then factor the equation as follows:

$$\sin^2\theta \, \sec \theta = \sec \theta$$

$$\sec \theta - \sin^2\theta \, \sec = 0$$

$$\sec \theta \, (1 - \sin^2\theta) = 0$$

Set each factor equal to zero

$$1 - \sin^2\theta = 0$$

51

and $\sec \theta = 0$

Solving the first equation

$$1 - \sin^2\theta = 0$$

$$\sin^2\theta = 1$$

$$\sin \theta = \pm 1$$

then

$$\theta = \arcsin \pm 1$$

$$= 90° \text{ and } 180°$$

Solving the second equation

$$\sec \theta = 0$$

then

$$\theta = \text{arc} \sec 0$$

However, there is no angle for which $\sec \theta$ equals zero and we reject this false solution. Therefore, the primary solutions for the original equation are 90° and 180°.

The second type of limited solutions occur when introducing a radical into an equation by substitution or by squaring both members of an equation in solving the equation. These solutions are called extraneous roots. Solutions of this type must be substituted into the original equation for verification.

EXAMPLE: Find the primary values of θ for the equation

$$\tan \theta - \sec \theta + 1 = 0$$

SOLUTION: We first rearrange the equation to read

$$\tan \theta + 1 = \sec \theta$$

Square both sides:

$$\tan^2\theta + 2 \tan \theta + 1 = \sec^2\theta$$

Rearrange again:

$$2 \tan \theta = \sec^2\theta - (\tan^2\theta + 1)$$

This gives

$$2 \tan \theta = 0$$

and

$$\tan \theta = 0$$

Therefore,

$$\theta = \arctan 0$$
$$= 0° \text{ and } 180°$$

These seem to be the values of the equation, but we squared both sides of the equation and we must now substitute these values into the original equation to verify the values. Upon substituting we find

$$\tan \theta - \sec \theta + 1 = 0$$

This implies that

$$\tan 0° - \sec 0° + 1 = 0$$

and since

$$0 - 1 + 1 = 0$$

the value 0° holds true.

For 180° we find

$$\tan \theta - \sec \theta + 1 = 0$$

53

$$\tan 180° - \sec 180° + 1 = 0$$

but $\qquad 1 - (-1) + 1 \neq 0$

and we say $180°$ is an extraneous root.

Practice Problems

Find the primary values for θ in the following equations:

1. $\tan \theta \cos^2 \theta = \sin^2 \theta$

2. $\cos^2 \theta \sin \theta = \sin \theta + 1$

3. $2 \sec \theta + 1 - \cos \theta = 0$

4. $\cot \theta - \csc \theta - \sqrt{3} = 0$

Answers

1. $0°$, $45°$, $180°$, $225°$

2. $270°$

3. $180°$

4. $240°$

Chapter 9
Straight Lines

The study of straight lines provides an excellent introduction to analytic geometry. As its name implies, this branch of mathematics is concerned with geometrical relationships. However, in contrast to plane and solid geometry, the study of these relationships in analytic geometry is accomplished by algebraic analysis.

The invention of the rectangular coordinate system made algebraic analysis of geometrical relationships possible. Rene Descartes, a French mathematician, is credited with this invention, and the coordinate system is often designated as the Cartesian coordinate system in his honor.

Recalling our previous study of the rectangular coordinate system, we review the following definitions and terms:

1. Distances measured along, or parallel to, the X axis are ABSCISSAS. They are positive if measured to the right of the origin; they are negative if measured to the left of the origin. (See fig. 9-1.)

2. Distances measured along, or parallel to, the Y axis are ORDINATES. They are positive

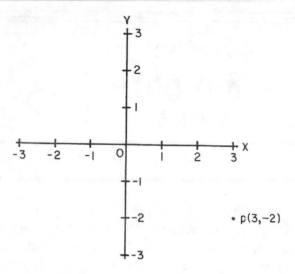

Figure 9-1. – Rectangular coordinate system.

if measured above the origin; they are negative if measured below the origin.

3. Any point on the coordinate system is designated by naming its abscissa and ordinate. For example, the abscissa of point P (fig. 9-1) is 3 and the ordinate is -2. Therefore, the symbolic notation for P is

$$P(3, -2)$$

In using this symbol to designate a point, the abscissa is always written first, followed by a comma. The ordinate is written last. Thus the general form of the symbol is

$$P(x, y)$$

4. The abscissa and ordinate of a point are its COORDINATES.

Distance Between Two Points

The distance between two points, P_1 and P_2, can be expressed in terms of their coordinates by using the Pythagorean Theorem. From our study of Mathematics, Vol. 1, NavPers 10069-C, we recall that this theorem is stated as follows:

In a right triangle, the square of the length of the hypotenuse (longest side) is equal to the sum of the squares of the lengths of the two shorter sides.

Let the coordinates of P_1 be (x_1, y_1) and let those of P_2 be (x_2, y_2), as in figure 9-2. By the Pythagorean Theorem,

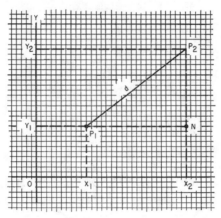

Figure 9-2. – Distance between two points.

$$d = \sqrt{(P_1 N)^2 + (P_2 N)^2}$$

where d represents the distance from P_1 to P_2. We can express the length of $P_1 N$ in terms of x_1 and x_2 as follows:

$$P_1 N = x_2 - x_1$$

Likewise, $\qquad P_2N = y_2 - y_1$

By substitution in the formula developed previously for d, we reach the following conclusion:

$$d = \sqrt{(x_2 - x_1)^2 + (y_2 - y_1)^2}$$

Although we have demonstrated the formula for the first quadrant only, it can be proved for all quadrants and all pairs of points.

EXAMPLE: In figure 9-2, $x_1 = 2$, $x_2 = 6$, $y_1 = 2$, and $y_2 = 5$. Find the length of d.

$$\text{SOLUTION: } d = \sqrt{(6 - 2)^2 + (5 - 2)^2}$$

$$= \sqrt{4^2 + 3^2}$$

$$= \sqrt{16 + 9}$$

$$= \sqrt{25}$$

$$= 5$$

This result could have been foreseen by observing that triangle P_1NP_2 is a 3-4-5 triangle.

EXAMPLE: Find the distance between P_1 (4, 6) and P_2(10, 4).

$$\text{SOLUTION: } d = \sqrt{(10 - 4)^2 + (4 - 6)^2}$$

$$d = \sqrt{36 + 4}$$

$$d = 2\sqrt{10}$$

Division of a Line Segment

Many times it becomes necessary to find the coordinates of a point which is some known fraction of the distance between P_1 and P_2.

In figure 9-3, P is a point lying on the line joining P_1 and P_2 so that

$$\frac{P_1P}{P_1P_2} = k$$

If P should lie one-quarter of the way between P_1 and P_2, then k would equal 1/4.

Triangles P_1MP and P_1NP_2 are similar. Therefore,

$$\frac{P_1M}{P_1N} = \frac{P_1P}{P_1P_2}$$

Since $\dfrac{P_1P}{P_1P_2}$ is the ratio that defines k,

$$\frac{P_1M}{P_1N} = k$$

Therefore,

$$P_1M = k\,(P_1N)$$

Referring again to figure 9-3, observe that P_1N is equal to $x_2 - x_1$. Likewise, P_1M is equal to $x - x_1$. Therefore, replacing P_1M and P_1N with their equivalents in terms of x, the foregoing equation becomes

$$x - x_1 = k(x_2 - x_1)$$
$$x = x_1 + k(x_2 - x_1)$$

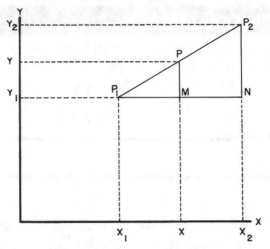

Figure 9-3. – Division of a Line Segment.

By similar reasoning, $y = y_1 + k(y_2 - y_1)$

The x and y found as a result of the foregoing discussion are the coordinates of the desired point, whose distances from P_1 and from P_2 are determined by the value of k.

EXAMPLE: Find the coordinates of a point 1/4 of the way from P_1 (2,3) to P_2 (4,1).

SOLUTION:

$$k = \frac{1}{4} \, , \, x_2 - x_1 = 2, \, y_2 - y_1 = -2$$

$$x = 2 + \frac{1}{4}(2) = 2 + \frac{1}{2} = \frac{5}{2}$$

$$y = 3 + \frac{1}{4}(-2) = 3 - \frac{1}{2} = \frac{5}{2}$$

The point P is $(\frac{5}{2}, \frac{5}{2})$.

Finding the Midpoint

When the midpoint of a line segment is to be found, the value of k is $1/2$. Therefore,

$$x = x_1 + \frac{1}{2} (x_2 - x_1)$$

$$= x_1 + \frac{1}{2} x_2 - \frac{1}{2} x_1$$

$$= \frac{1}{2} x_1 + \frac{1}{2} x_2$$

$$= \frac{1}{2} (x_1 + x_2)$$

By similar reasoning,

$$y = \frac{1}{2} (y_1 + y_2)$$

EXAMPLE: Find the midpoint of the line between P_1 (2,4) and P_2 (4,6).

SOLUTION: $k = \dfrac{1}{2}$

$$x = \frac{1}{2} (2 + 4)$$

$$= 3$$

$$y = \frac{1}{2} (4 + 6)$$

$$= 5$$

The midpoint is (3,5).

Inclination and Slope

A line drawn on the rectangular coordinate system and crossing the X axis forms a positive acute angle with the X axis. This angle, shown in figure 9-4 as angle α, is called the angle of inclination of the line.

The slope of any line, such as AB in figure 9-4, is equal to the tangent of its angle of inclination. Slope is denoted by the letter m. Therefore, for line AB,

$$m = \tan \alpha$$

If the axes are in their conventional positions, a line sloping upward to the right has a positive slope. A line sloping downward to the right has a negative slope.

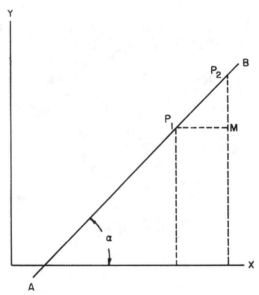

Figure 9-4. – Angle of Inclination.

Since the tangent of α is the ratio of P_2M to P_1M, we can relate the slope of line AB to the points P_1 and P_2 as follows:

$$m = \tan \alpha = \frac{P_2M}{P_1M}$$

Designating the coordinates of P_1 as (x_1, y_1), and those of P_2 as (x_2, y_2), we recall that

$$P_2M = y_2 - y_1$$

$$P_1M = x_2 - x_1$$

$$m = \frac{y_2 - y_1}{x_2 - x_1}$$

The quantities $(x_2 - x_1)$ and $(y_2 - y_1)$ represent changes that occur in the values of the x and y coordinates as a result of changing from P_2 to P_1 on line AB. The symbol used by mathematicians to represent an increment of change is the Greek letter delta (Δ). Therefore, Δx means "the change in x" and Δy means "the change in y." The amount of change in the x coordinate, as we change from P_2 to P_1, is $x_2 - x_1$. Therefore,

$$\Delta x = x_2 - x_1$$

$$\Delta y = y_2 - y_1$$

We use this notation to express the slope of line AB, as follows:

$$m = \frac{\Delta y}{\Delta x}$$

EXAMPLE: Find the slope of the line connecting P_2 (7,6) and P_1 (-1,-4).

SOLUTION:

$$m = \frac{\Delta y}{\Delta x}$$

$$\Delta y = y_2 - y_1 = 6 - (-4) = 10$$

$$\Delta x = x_2 - x_1 = 7 - (-1) = 8$$

$$m = \frac{10}{8} = \frac{5}{4}$$

It is important to realize that the choice of labels for P_1 and P_2 is strictly arbitrary. If we had chosen the point (7,6) to be P_1 in the foregoing example, and the point (-1,-4) to be P_2, the following calculation would have resulted:

$$m = \frac{\Delta y}{\Delta x}$$

$$\Delta y = y_2 - y_1 = -4 - 6 = -10$$

$$\Delta x = x_2 - x_1 = -1 - 7 = -8$$

$$m = \frac{-10}{-8} = \frac{5}{4}$$

This is the same result as in the foregoing example.

The slope of $5/4$ means that a point moving along this line would move vertically $+5$ units for every horizontal movement of $+4$ units. This result is consistent with the previously stated meaning of positive slope; i.e., sloping upward to the right.

If line AB in figure 9-4 were parallel to the X axis, y_1 and y_2 would be equal and the difference $(y_2 - y_1)$ would be 0. Therefore,

$$m = \frac{0}{x_2 - x_1} = 0$$

Thus we conclude that the slope of a horizontal line is 0. This conclusion can also be reached by noting that angle α (fig. 9-4) is 0 when the line is parallel to the X axis. Since the tangent of $0°$ is 0,

$$m = \tan \alpha = 0$$

The slope of a line that is parallel to the Y axis becomes meaningless. The tangent of the angle α increases indefinitely as α approaches $90°$. It is sometimes said that $m \to \infty$ (m approaches infinity) when α approaches $90°$.

Parallel and Perpendicular Lines

If we are given two lines that are parallel, their slopes must be equal. Each line will cut the X axis at the same angle α, so that

$$m_1 = \tan \alpha, \; m_2 = \tan \alpha$$

Therefore, $\qquad m_1 = m_2$

We conclude that two lines which are parallel have the same slope.

Suppose that two lines are perpendicular to each other, as lines L_1 and L_2 in figure 9-5. The slope and inclination of L_1 are m_1 and α_1, respectively. The slope and inclination of L_2 are m_2 and α_2, respectively. Then the following is true:

$$m_1 = \tan \alpha_1$$

$$m_2 = \tan \alpha_2$$

It can be shown geometrically that α_2 (fig. 9-5) is equal to α_1, plus $90°$. Therefore,

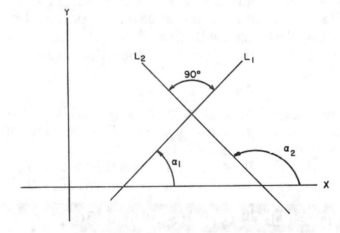

Figure 9-5. – Slopes of perpendicular lines.

$$\tan \alpha_2 = \tan (\alpha_1 + 90°)$$

$$= - \cot \alpha_1$$

$$= - \frac{1}{\tan \alpha_1}$$

Replacing tan α_1 and tan α_2 by their equivalents in terms of slope, we have

$$m_2 = -\frac{1}{m_1}$$

We conclude that, if two lines are perpendicular, the slope of one is the negative reciprocal of the slope of the other.

Conversely, if the slopes of two lines are negative reciprocals of each other, the lines are perpendicular.

EXAMPLE: In figure 9-6, show that line L_1 is perpendicular to line L_2. Line L_1 passes through points P_1 (0,5) and P_2 (-1,3). Line L_2 passes through points P_2 (-1,3) and P_3 (3,1).

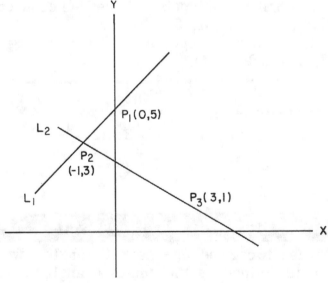

Figure 9-6. – Proving lines perpendicular.

SOLUTION: Let m_1 and m_2 represent the slopes of lines L_1 and L_2, respectively. Then we have

$$m_1 = \frac{5 - 3}{0 - (-1)} = 2$$

$$m_2 = \frac{1 - 3}{3 - (-1)} = \frac{-2}{4} = -\frac{1}{2}$$

Since their slopes are negative reciprocals of each other, the lines are perpendicular.

Practice Problems

1. Find the distance between P_1 (5,3) and P_2 (6,7).
2. Find the distance between P_1 (1/2,1) and P_2 (3/2,5/3).
3. Find the midpoint of the line connecting P_1 (5,2) and P_2 (-1,-3).
4. Find the slope of the line joining P_1 (-2,-5) and P_2 (2,5).

Answers

1. $\sqrt{17}$

2. $\dfrac{\sqrt{13}}{3}$

3. $(2, -\dfrac{1}{2})$

4. $\dfrac{5}{2}$

Angle Between Two Lines

When two lines intersect, the angle between them is defined as the smallest angle through which one of the lines must be rotated to make it coincide with the other line. For example, the angle ϕ in figure 9-7 is the angle between lines L1 and L2.

Referring to figure 9-7,

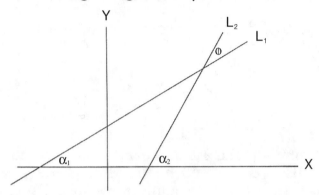

Figure 9-7. – Angle Between Two Lines

$$\alpha_2 = \alpha_1 + \phi$$

$$\therefore \phi = \alpha_2 - \alpha_1$$

It is possible to determine the value of ϕ directly from the slopes of lines L1 and L2, as follows:

$$\tan \phi = \tan (\alpha_2 - \alpha_1)$$

$$= \frac{\tan \alpha_2 - \tan \alpha_1}{1 + \tan \alpha_1 \tan \alpha_2}$$

This result is obtained by use of the trigonometric identity for the tangent of the difference between two angles. Trigonometric identities are discussed in chapter 8 of this training course.

Recalling that the tangent of the angle of inclination is the slope of the line, we have

$$\tan \alpha_1 = m_1 \text{ (the slope of } L_1)$$

$$\tan \alpha_2 = m_2 \text{ (the slope of } L_2)$$

Substituting these expressions in the tangent formula derived in the foregoing discussion, we have

$$\tan \phi = \frac{m_2 - m_1}{1 + m_1 m_2}$$

If one of the lines were parallel to the Y axis, its slope would be infinite. This would render the slope formula for $\tan \phi$ useless, because an infinite value in both the numerator and denominator of the fraction $\frac{m_2 - m_1}{1 + m_1 m_2}$ produces an indeterminate form. However, if one of the lines is known to be parallel to the Y axis the tangent of ϕ may be expressed by another method.

Suppose that L_2 (fig. 9-7) were parallel to the Y axis. Then we would have

$$\alpha_2 = 90°$$

$$\phi = 90° - \alpha_1$$

$$\tan \phi = \cot \alpha_1$$

$$= \frac{1}{m_1}$$

70

Practice Problems

1. Find the angle between the two lines which have $m_1 = 3$ and $m_2 = 7$ for slopes.

2. Find the angle between two lines whose slopes are $m_1 = 0$, $m_2 = 1$. ($m_1 = 0$ signifies that line L_1 is horizontal and the formula still holds).

3. Find the angle between the Y axis and a line with a slope of $m = -8$.

4. Find the obtuse angle between the X axis and line with a slope of $m = -8$.

Answers

1. $10°18'$ 3. $7°7'$
2. $45°$ 4. $97°7'$

Equation of a Straight Line

Equations such as $2x + y = 6$
are designated as linear equations, and their graphs are shown to be straight lines. The purpose of the present discussion is to study the relationship of slope to the equation of a straight line.

Point-Slope Form

Suppose that we desire to find the equation of a straight line which passes through a known point and has a known slope. Let (x,y) represent the coordinates of any point on the line, and let (x_1,y_1) represent the coordinates of the known point. The slope is represented by m.

Recalling the formula defining slope in terms of the coordinates of two points, we have

$$m = \frac{y - y_1}{x - x_1}$$

$$\therefore y - y_1 = m(x - x_1)$$

EXAMPLE: Find the equation of a line passing through the point (2,3) and having a slope of 3.

SOLUTION:

$$x_1 = 2 \text{ and } y_1 = 3$$

$$y - y_1 = m(x - x_1)$$

$$\therefore y - 3 = 3(x - 2)$$

$$y - 3 = 3x - 6$$

$$y - 3x = -3$$

The point-slope form may be used to find the equation of a line through two known points. The values of x_1, x_2, y_1, and y_2 are first used to find the slope of the line, and then either known point is used with the slope in the point-slope form.

EXAMPLE: Find the equation of the line through the points (-3,4) and (4,-2).

SOLUTION:

$$m = \frac{y_2 - y_1}{x_2 - x_1}$$

$$= \frac{-2 - 4}{4 + 3} = \frac{-6}{7}$$

Letting (x,y) represent any point on the line, and using (-3,4) as a known point, we have

$$y - 4 = \frac{-6}{7} \left[x - (-3) \right]$$
$$7(y - 4) = -6(x + 3)$$
$$7y - 28 = -6x - 18$$
$$7y + 6x = 10$$

Slope-Intercept Form

Any line which is not parallel to the Y axis intersects the Y axis in some point. The x coordinate of the point of intersection is 0, because the Y axis is vertical and passes through the origin. Let the y coordinate of the point of intersection be represented by b. Then the point of intersection is (0,b), as shown in figure 9-8. The y coordinate, b, is called the y intercept.

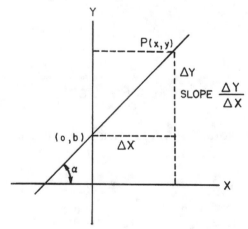

Figure 9-8. – Slope-Intercept Form.

The slope of the line in figure 9-8 is $\frac{\Delta y}{\Delta x}$. The value of Δy in this expression is $y - b$, where y represents the y coordinate of any point on the line. The value of Δx is equal to the x coordinate of $P(x,y)$, so that

$$m = \frac{\Delta y}{\Delta x} = \frac{y - b}{x}$$

$$mx = y - b$$

$$y = mx + b$$

This is the standard slope-intercept form of a straight line.

EXAMPLE: Find the equation of a line that intersects the Y axis at the point (0,3) and has a slope of 5/3.

SOLUTION:

$$y = mx + b$$

$$y = \frac{5}{3} x + 3$$

$$3y = 5x + 9$$

Practice Problems

Write equations for lines having points and slopes as follows:

1. $P(3,5)$, $m = -2$

2. $P(-2,-1)$, $m = \frac{1}{3}$

3. $P_1(2,2)$ and $P_2(-4,-1)$

4. Y intercept = 2, $m = 3$

Answers

1. $y = -2x + 11$

2. $3y = x - 1$

3. $2y = x + 2$

4. $y = 3x + 2$

Normal Form

Methods for determining the equation of a line usually depend upon some knowledge of a point or points on the line. We now consider a method which does not require advance knowledge concerning any of the line's points. All that is known about the line is its perpendicular distance from the origin and the angle between the perpendicular and the x axis.

In figure 9-9, line AB is a distance p away from the origin, and line OM forms an angle θ with the X axis. We select any point $P(x,y)$ on line AB and develop the equation of line AB in terms of the x and y of P. Since P represents ANY point on the line, the x and y of the equation will represent EVERY point on the line and therefore will represent the line itself.

PR is constructed perpendicular to OB at point R. NR is drawn parallel to AB, and PN is parallel to OB. PS is perpendicular to NR and to AB. Since right triangles OMB and RSP have their sides mutually perpendicular, they are similar; therefore, angle PRS is equal to θ. Finally, the x distance of point P is equal to OR, and the y distance of P is equal to PR.

In order to relate the distance p to x and y, we reason as follows:

$$ON = (OR)(\cos\theta)$$

$$= x\cos\theta$$

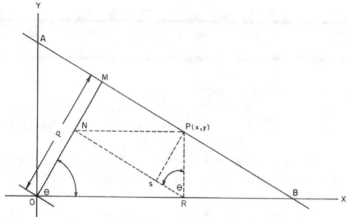

Figure 9-9. – Normal Form.

$$PS = (PR)(\sin\theta)$$
$$= y\sin\theta$$
$$OM = ON + PS$$
$$p = ON + PS$$
$$p = x\cos\theta + y\sin\theta$$

This final equation is the NORMAL FORM. The word "normal" in this usage refers to the perpendicular relationship between OM and AB. "Normal" frequently means "perpendicular" in mathematical and scientific usage. The distance p is considered to be always positive, and θ is any angle between 0° and 360°.

EXAMPLE: Find the equation of a line that is 5 units away from the origin, if the perpendicular from the line to the origin forms an angle of 30° with the positive side of the X axis.

SOLUTION:

$$p = 5;\ \theta = 30°$$

$$p = x \cos \theta + y \sin \theta$$

$$5 = x \cos 30° + y \sin 30°$$

$$5 = x \left(\frac{\sqrt{3}}{2}\right) + y \left(\frac{1}{2}\right)$$

$$10 = x \sqrt{3} + y$$

Parallel and Perpendicular Lines

The general equation of a straight line is often written with capital letters for co-efficients, as follows:

$$Ax + By + C = 0$$

These literal coefficients, as they are called, represent the numerical coefficients encountered in a typical linear equation.

Suppose that we are given two equations which are duplicates except for the constant term, as follows:

$$Ax + By + C = 0$$

$$Ax + By + D = 0$$

By placing these two equations in slope-intercept form, we can show that their slopes are equal, as follows:

$$y = \left(- \frac{A}{B}\right) x + \left(- \frac{C}{B}\right)$$

$$y = \left(- \frac{A}{B}\right) x + \left(- \frac{D}{B}\right)$$

Thus the slope of each line is $- A/B$.

Since lines having equal slopes are parallel, we reach the following conclusion: In any two linear equations, if the coefficients of the x and y terms are identical in value and sign, then the lines represented by these equations are parallel.

EXAMPLE: Write the equation of a line parallel to $3x - y - 2 = 0$ and passing through the point (5,2).

SOLUTION: The coefficients of x and y in the desired equation are the same as those in the given equation. Therefore, the equation is

$$3x - y + D = 0$$

Since the line passes through (5,2), the values $x = 5$ and $y = 2$ must satisfy the equation. Substituting these, we have

$$3(5) - (2) + D = 0$$

$$D = -13$$

Thus the required equation is

$$3x - y - 13 = 0$$

A situation similar to that prevailing with parallel lines involves perpendicular lines. For example, consider the equations

78

$$Ax + By + C = 0$$

$$Bx - Ay + D = 0$$

Transposing into the slope-intercept form, we have

$$y = \left(-\frac{A}{B}\right) x + \left(-\frac{C}{B}\right)$$

$$y = \left(\frac{B}{A}\right) x + \left(\frac{D}{A}\right)$$

Since the slopes of these two lines are negative reciprocals, the lines are perpendicular.

The conclusion derived from the foregoing discussion is as follows: If a line is to be perpendicular to a given line, the coefficients of x and y in the required equation are found by interchanging the coefficients of x and y in the given equation and changing the sign of one of them.

EXAMPLE: Write the equation of a line perpendicular to the line $x + 3y + 3 = 0$ and having a y intercept of 5.

SOLUTION: The required equation is

$$3x - y + D = 0$$

Notice the interchange of coefficients and the change of sign. At the point where the line crosses the Y axis, the value of x is 0 and the value of y is 5. Therefore, the equation is

$$3(0) - (5) + D = 0$$

$$D = 5$$

The required equation is

$$3x - y + 5 = 0$$

Practice Problems

Find the equations of the following lines:

1. Through (1,1) and parallel to 5x - 3y = 9.

2. Through (-3,2) and perpendicular to x + y = 5.

3. Through (2,3) and perpendicular to 3x - 2y = 7.

4. Through (2,3) and parallel to 3x - 2y = 7.

Answers

1. 5x - 3y = 2

2. x - y = -5

3. 2x + 3y = 13

4. 3x - 2y = 0

Distance of a Point from a Line

It is frequently necessary to express the distance of a point from a line in terms of the coefficients in the equation of the line. In order to do this, we compare the two forms of the equation of a straight line, as follows:

General equation: Ax + By + C = 0
Normal form: x cos θ + y sin θ - p = 0

The general equation and the normal form represent the same straight line. Therefore, A (the coefficient of x in general form) is proportional to cos θ (the coefficient of x in the normal form). By similar reasoning, B is proportional to sin θ , and C is proportional to -p. Recalling that quantities proportional to each other form ratios involving a constant of proportionality, let k be this constant. Thus we have

$$\frac{\cos \theta}{A} = k$$

$$\frac{\sin \theta}{B} = k$$

$$\cos \theta = kA$$

$$\sin \theta = kB$$

Squaring both sides of these two expressions and then adding, we have

$$\cos^2 \theta + \sin^2 \theta = k^2 (A^2 + B^2)$$

$$\therefore 1 = k^2 (A^2 + B^2)$$

$$k^2 = \frac{1}{A^2 + B^2}$$

$$k = \frac{1}{\pm \sqrt{A^2 + B^2}}$$

The coefficients in the normal form, expressed in terms of A, B, and C, are as follows:

$$\cos \theta = \frac{A}{\pm \sqrt{A^2 + B^2}}$$

$$\sin \theta = \frac{B}{\pm \sqrt{A^2 + B^2}}$$

$$-p = \frac{C}{\pm \sqrt{A^2 + B^2}}$$

The sign of $\sqrt{A^2 + B^2}$ is chosen so as to make p (a distance) always positive.

The conversion formulas developed in the foregoing discussion are used in finding the distance from a point to a line. Let p represent the distance of line L from the origin. (See fig. 9-10.) In order to find d, the distance

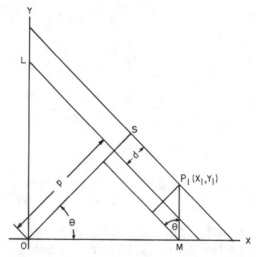

Figure 9-10. – Distance from a point to a line.

of point P_1 from line L, we construct a line through P_1 and parallel to L. The distance of this line from the origin is OS, and the difference between OS and p is d.

We obtain an expression for d, based on the coordinates of P_1, as follows:

$$OS = x_1 \cos \theta + y_1 \sin \theta$$

$$d = OS - p$$

$$= x_1 \cos \theta + y_1 \sin \theta - p$$

Returning to the expressions for $\sin \theta$, $\cos \theta$, and -p in terms of A, B, and C (the coefficients in the general equation), we have

$$d = x_1 \left(\frac{A}{\pm \sqrt{A^2 + B^2}} \right) + y_1 \left(\frac{B}{\pm \sqrt{A^2 + B^2}} \right)$$

$$+ \frac{C}{\pm \sqrt{A^2 + B^2}}$$

The denominator in each of the expressions comprising the formula for d is the same. Therefore we may combine as follows:

$$d = \left| \frac{x_1 A + y_1 B + C}{\sqrt{A^2 + B^2}} \right|$$

We use the absolute value, since d is a distance, and thus avoid any confusion arising from the ± radical.

EXAMPLE: Find the distance from the point (2,1) to the line $4x + 2y + 7 = 0$.

83

SOLUTION:

$$d = \left| \frac{(4)\ (2)\ +\ (2)\ (1)\ +\ 7}{\sqrt{4^2 + 2^2}} \right|$$

$$= \frac{8 + 2 + 7}{\sqrt{20}}$$

$$= \frac{17}{2\sqrt{5}}$$

$$= \frac{17\sqrt{5}}{10}$$

Practice Problems

In each of the following problems, find the distance from the point to the line:

1. $(5,2)$, $3x - y + 6 = 0$

2. $(3,-5)$, $2x + y + 4 = 0$

3. $(3,-4)$, $4x + 3y = 10$

4. $(-2,5)$, $3x + 4y - 9 = 0$

Answers

1. $\dfrac{19\sqrt{10}}{10}$ 3. 2

2. $\sqrt{5}$ 4. 1

Chapter 10
Conic Sections

This chapter is a continuation of the study of analytic geometry. The figures presented in this chapter are plane figures which are included in the general class of conic sections or simply "conics."

Conic sections are so named because they are all plane sections of a right circular cone. A circle can be formed by cutting a cone perpendicular to its axis. An ellipse is produced when the cone is cut obliquely to the axis and the surface. A hyperbola results when the cone is intersected by a plane parallel to the axis, and a parabola is the result when the intersecting plane is parallel to an element of the surface. These are illustrated in figure 10-1.

When the curve produced by cutting the cone is placed on a coordinate system it may be defined as follows:

A conic section is the locus of a point that moves so that its distance from a fixed point is in a constant ratio to its distance from a fixed line. The fixed point is the focus, and the fixed line is the directrix.

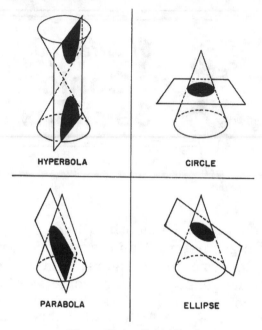

HYPERBOLA   CIRCLE

PARABOLA   ELLIPSE

Figure 10-1 – Conic Sections.

The ratio referred to in the definition is called the eccentricity. If the eccentricity (e) is less than one, the curve is an ellipse. If e is greater than one, the curve is a hyperbola. If e is equal to 1, the curve is a parabola. A circle is a special case having an eccentricity equal to zero, and may be defined by the distance from a point. It is actually a limiting case of an ellipse in which the eccentricity approaches zero. Thus, if

$$e = 0, \text{ it is a circle}$$
$$e < 1, \text{ it is an ellipse}$$
$$e = 1, \text{ it is a parabola}$$
$$e > 1, \text{ it is a hyperbola}$$

The eccentricity, focus, and directrix are used in the algebraic analysis of conic sections and the corresponding equations. The concept of the locus of an equation also enters into analytic geometry; this concept is discussed before the individual conic sections are studied.

The Locus of an Equation

In chapter 9 of this course, methods for analysis of linear equations are presented. If a group of x and y values (or ordered pairs, P (x,y)) which satisfy a given linear equation are plotted on a coordinate system, the resulting graph is a straight line.

When higher ordered equations such as

$$x^2 + y^2 = 1 \text{ or } y = \sqrt{2x + 3}$$

are encountered, the resulting graph is not a straight line. However, the points whose coordinates satisfy most of the equations in x and y are normally not scattered in a random field. If the values are plotted they will seem to follow a line or curve (or a combination of lines and curves). In many texts the plot of an equation is called a curve, even when it is a straight line. This curve is called the locus of the equation. The locus of an equation is a curve containing those points, and only those points, whose coordinates satisfy the equation.

At times the curve may be defined by a set of conditions rather than by an equation, though an equation may be derived from the given conditions. Then the curve in question would be the locus of all points which fit the conditions. For instance a circle may be said to be the locus of

87

all points in a plane which lie a fixed distance from a fixed point. A straight line may be defined as the locus of a point that moves in a plane so that it is at all times equidistant from two fixed points. The method of expressing a set of conditions in analytical form gives an equation. Let us draw up a set of conditions and translate them into an equation.

EXAMPLE: What is the equation of the curve which is the locus of all points which are equidistant from the two points (5, 3) and (2,1)?

SOLUTION: First, as in figure 10-2, choose some point having coordinates (x, y). Recall from chapter 9 of this course that the distance between this point and (2, 1) is given by:

$$\sqrt{(y - 1)^2 + (x - 2)^2}$$

The distance between point (x, y) and (5, 3) will be given by

$$\sqrt{(y - 3)^2 + (x - 5)^2}$$

Equating these distances, since the point is to be equidistant from the two given points, we have

$$\sqrt{(y - 1)^2 + (x - 2)^2} = \sqrt{(y - 3)^2 + (x - 5)^2}$$

Squaring both sides

$$(y - 1)^2 + (x - 2)^2 = (y - 3)^2 + (x - 5)^2$$

Expanding

$$y^2 - 2y + 1 + x^2 - 4x + 4$$
$$= y^2 - 6y + 9 + x^2 - 10x + 25$$

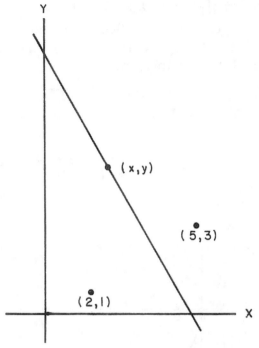

Figure 10-2. – Locus of points equidistant from two given points..

Canceling and collecting terms:

$$4y + 5 = -6x + 34$$

$$4y = -6x + 29$$

$$y = - \frac{3}{2}x + 7.25$$

This is the equation of a straight line with a slope of minus $3/2$, and a Y intercept of $+7.25$.

EXAMPLE: Find the equation of the curve which is the locus of all points which are equidistant from the line $x = -3$ and the point $(3, 0)$.

SOLUTION: The distance from the point (x, y) on the curve to the line will be (x - (-3)) or (x + 3). Refer to figure 10-3. The distance from the point (x, y) to the point (3, 0) is

$$\sqrt{(y - 0)^2 + (x - 3)^2}$$

Equating the two distances,

$$x + 3 = \sqrt{y^2 + (x - 3)^2}$$

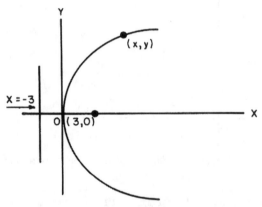

Figure 10-3. – Parabola.

Squaring both sides,

$$x^2 + 6x + 9 = y^2 + x^2 - 6x + 9$$

Canceling and collecting terms,

$$y^2 = 12x$$

which is the equation of a parabola.

EXAMPLE: What is the equation of the curve the locus of which is a point which moves so that at all times the ratio of its distance from the point (3, 0) to its distance from the line x = 25/3 is equal to 3/5? Refer to figure 10-4.

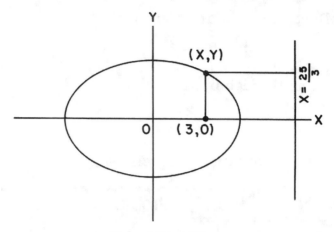

Figure 10-4. – Ellipse.

SOLUTION: The distance from a point (x, y) to the point $(3, 0)$ is given by

$$d_1 = \sqrt{(x - 3)^2 + (y - 0)^2}$$

The distance from the same point (x, y) to the line is

$$d_2 = \frac{25}{3} - x$$

Since $\quad\dfrac{d_1}{d_2} = \dfrac{3}{5}$ or $d_1 = \dfrac{3}{5} d_2$

then $\quad\sqrt{(x - 3)^2 + y^2} = \dfrac{3}{5}\left(\dfrac{25}{3} - x\right)$

Squaring both sides and expanding,

$$x^2 - 6x + 9 + y^2 = \frac{9}{25}\left(x^2 - \frac{50}{3}x + \frac{625}{9}\right)$$

$$x^2 - 6x + 9 + y^2 = \frac{9}{25}x^2 - 6x + 25$$

Collecting terms and transposing

$$\frac{16}{25} x^2 + y^2 = 16$$

Dividing through by 16

$$\frac{x^2}{25} + \frac{y^2}{16} = 1$$

This is the equation of an ellipse.

Practice Problems

Find the equation which is the locus of the point which moves so that it is at all times:

1. Equidistant from the points (0, 0) and (5, 4).

2. Equidistant from the points (3, -2) and (-3, 2).

3. Equidistant from the line x = -4 and the point (3, 4).

4. Equidistant from the point (4, 5) and the line y = 5x -4. HINT: Use the standard distance formula to find the distance from the point P (x, y) and the point P (4, 5). Then use the formula for finding distance from a point to a line, given in chapter 9 of this course, to find the distance from P (x, y) to the given line. Put the equation of the line in the form Ax + By + C = 0.

Answers

1. $y = - 1.25x + \frac{41}{8}$

2. $2y = 3x$

3. $y^2 - 8y = 14x - 9$

4. $x^2 + 10xy + 25y^2 - 168x - 268y + 1050 = 0$

The Circle

A circle is the locus of a point which is always a fixed distance from a fixed point called the center.

The fixed distance spoken of here is the radius of the circle.

The equation of a circle with its center at the origin (figure. 10-5) is, from the definition:

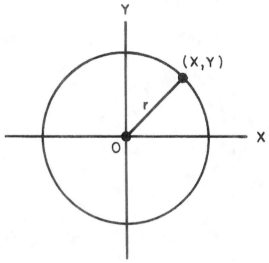

Figure 10-5. – Circle with center at the origin.

$$\sqrt{(x - 0)^2 + (y - 0)^2} = r,$$

where (x, y) is a point on the circle and r is the radius and replaces d in the standard distance formula. Then

$$\sqrt{x^2 + y^2} = r$$

or $\qquad\qquad\qquad\qquad\qquad\qquad\qquad\qquad$ (1)

$$x^2 + y^2 = r^2$$

If the center of a circle, figure 10-6, is at some point $x = h$, $y = k$, the distance of the moving point from the center will be constant and equal to

$$\sqrt{(x - h)^2 + (y - k)^2} = r$$

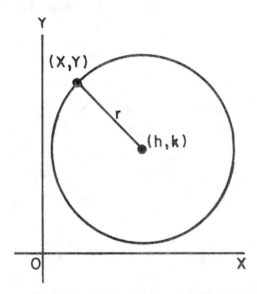

Figure 10-6. – Circle with center at (h, k).

or

$$(x - h)^2 + (y - k)^2 = r^2 \qquad\qquad (2)$$

Equations (1) and (2) are the standard forms for the equation of a circle. Equation (1) is merely

94

a special case of equation (2) in which h and k are equal to zero.

The equation of a circle may also be expressed in the form:

$$x^2 + y^2 + Bx + Cy + D = 0 \qquad (3)$$

where B, C, and D are constants.

THEOREM: An equation of the second degree in which the coefficients of the x^2 and y^2 terms are equal, and there is no (xy) term, represents a circle.

Whenever we find an equation in the form of equation (3), it is best to convert it to the form of equation (2), so that we have the coordinates of the center of the circle and the radius as part of the equation. This may be done as shown in the following example problems.

EXAMPLE: Find the coordinates of the center and the radius of the circle which is described by the following equation:

$$x^2 + y^2 - 4x - 6y + 9 = 0$$

SOLUTION: First rearrange the terms

$$x^2 - 4x + y^2 - 6y + 9 = 0$$

and complete the square in both x and y. Completing the square is discussed in a previous chapter on quadratic solutions. The procedure consists basically of adding certain quantities to both sides of a second degree equation to form a perfect square trinomial. When both the first and second degree members are known, the square of one half the coefficient of the first degree term is added to both sides of the equation. This will allow the quadratic equation to be factored into a perfect square trinomial. To complete the square in x in the given equation

$$x^2 - 4x + y^2 - 6y + 9 = 0$$

add the square of one-half the coefficient of x to both sides of the equation

$$x^2 - 4x + (2)^2 + y^2 - 6y + 9 = 0 + (2)^2$$

then $(x^2 - 4x + 4) + y^2 - 6y + 9 = 4$

$$(x-2)^2 + y^2 - 6y + 9 = 4$$

completes the square in x.
For y then

$$(x - 2)^2 + y^2 - 6y + (3)^2 + 9 = 4 + (3)^2$$

$$(x - 2)^2 + (y^2 - 6y + 9) + 9 = 4 + 9$$

$$(x - 2)^2 + (y - 3)^2 + 9 = 4 + 9$$

completes the square in y.

Transpose all constant terms to the right-hand side and simplify

$$(x - 2)^2 + (y - 3)^2 = 4 + 9 - 9$$

$$(x - 2)^2 + (y - 3)^2 = 4$$

and the equation is in the standard form of equation (2). This represents a circle with the center at (2, 3) and with a radius equal to $\sqrt{4}$ or 2.

EXAMPLE: Find the coordinates of the center and the radius of the circle given by the equation

$$x^2 + y^2 + \frac{1}{2}x - 3y - \frac{27}{16} = 0$$

SOLUTION: Rearrange and complete the squares in x and y

$$x^2 + \frac{1}{2}x + y^2 - 3y - \frac{27}{16} = 0$$

$$(x^2 + \frac{1}{2}x + \frac{1}{16}) + (y^2 - 3y + \frac{9}{4}) - \frac{27}{16} = \frac{1}{16} + \frac{9}{4}$$

Transposing all constant terms to the right-hand side and adding,

$$(x^2 + \frac{1}{2}x + \frac{1}{16}) + (y^2 - 3y + \frac{9}{4}) = 4$$

Reducing to standard form

$$(x + \frac{1}{4})^2 + (y - \frac{3}{2})^2 = (2)^2$$

Thus, the equation represents a circle with its center at (-1/4, 3/2) and a radius equal to 2.

Practice Problems

Find the coordinates of the center and the radius for the circles described by the following equations.

1. $x^2 - \frac{4}{5}x + y^2 - 4y + \frac{29}{25} = 0$

2. $x^2 + 6x + y^2 - 14y = 23$

3. $x^2 - 14x + y^2 + 22y = -26$

4. $x^2 + y^2 + \frac{2}{5}x + \frac{2}{3}y = \frac{2}{25}$

5. $x^2 + y^2 - 1 = 0$

Answers

1. Center $(\frac{2}{5}, 2)$, radius $\sqrt{3}$

2. Center $(-3, 7)$, radius 9

3. Center $(7, -11)$, radius 12

4. Center $(-\frac{1}{5}, -\frac{1}{3})$, radius $\frac{2\sqrt{13}}{15}$

5. Center $(0, 0)$, radius 1

The Circle Defined by Three Points

In certain situations it is convenient to consider the following standard form of a circle

$$x^2 + y^2 + Bx + Cy + D = 0$$

as the equation of a circle in which the specific values of the constants B, C, and D are to be determined. In this problem the unknowns to be found are not x and y, but the values of the constants B, C, and D. The conditions which define the circle are used to form algebraic relationships between these constants. For example, if one of the conditions imposed on the circle is that it pass through the point $(3, 4)$ then the standard form is written with x and y replaced by 3 and 4 respectively; thus

$$x^2 + y^2 + Bx + Cy + D = 0$$

is rewritten as

$$(3)^2 + (4)^2 + B(3) + C(4) + D = 0$$

$$3B + 4C + D = -25$$

There are three independent constants in the equation of a circle; therefore, there must be three conditions given to define a circle. Each of these conditions will yield an equation with B, C, and D as the unknowns. These three equations are then solved simultaneously to determine the values of the constants which satisfy all of the equations. In an analysis, the number of independent constants in the general equation of a curve indicate how many conditions must be set before a curve can be completely defined. Also, the number of unknowns in an equation indicates the number of equations which must be solved simultaneously to find the values of the unknowns. For example, if B, C, and D are unknowns in an equation, three separate equations involving these variables are required for a solution.

A circle may be defined by three noncollinear points, that is, by three points which do not lie on a straight line. There is only one possible circle through any three noncollinear points. To find the equation of the circle determined by the three points substitute the x, y values of each of the given points into a general equation to form three equations with B, C, and D as the unknowns. These equations are then solved simultaneously to find the values of B, C, and D in the equation which satisfies the three given conditions.

The solution of simultaneous equations in two variables has been discussed previously. Systems involving three variables use an extension of the same principles, but with three equations instead of two. Step-by-step explanations of the solution will be given in the example problems.

EXAMPLE: Write the equation of the circle which passes through the points (2, 8), (5, 7), and (6, 6).

SOLUTION: The method used in this solution corresponds to the addition-subtraction method used for solution of equations in two variables. However, the method or combination of methods used will depend on a particular problem. No one method is best suited to all problems.

First, write a general equation of the form

$$x^2 + y^2 + Bx + Cy + D = 0$$

for each of the given points, substituting the given values for x and y and rearranging

For (2, 8) $4 + 64 + 2B + 8C + D = 0$
            $2B + 8C + D = -68$
For (5, 7) $25 + 49 + 5B + 7C + D = 0$
            $5B + 7C + D = -74$
For (6, 6) $36 + 36 + 6B + 6C + D = 0$
            $6B + 6C + D = -72$

To aid in the explanation we number the three resulting equations

$$2B + 8C + D = -68 \qquad (1)$$
$$5B + 7C + D = -74 \qquad (2)$$
$$6B + 6C + D = -72 \qquad (3)$$

The first step is to eliminate one of the unknowns and have two equations and two unknowns remain. The coefficient of D is the same in all three equations and is the one most easily eliminated by addition and subtraction. This is done in the following manner. Subtract (2) from (1)

$$2B + 8C + D = -68 \qquad (1)$$
$$5B + 7C + D = -74 \qquad (-)\ (2)$$
$$\overline{}$$
$$-3B + C = 6 \qquad (4)$$

Subtract (3) from (2)

$$5B + 7C + D = -74 \qquad (2)$$
$$6B + 6C + D = -72 \qquad (3)$$
$$\overline{}$$
$$-B + C = -2 \qquad (5)$$

This gives two equations, (4) and (5), in two unknowns which can be solved simultaneously. Since the coefficient of C is the same in both equations it is the most easily eliminated variable.

To eliminate C, subtract (4) from (5)

$$-B + C = -2 \qquad (5)$$
$$-3B + C = 6 \qquad (4)$$
$$\overline{}$$
$$2B = -8$$
$$B = -4 \qquad (6)$$

To find the value of C substitute the value found for B in (6) in equation (5)

$$-B + C = -2 \qquad (5)$$
$$-(-4) + C = -2$$
$$C = -6$$

Now the values of B and C can be substituted in any one of the original equations to determine the value of D.

101

If the values are substituted in (1)

$$2B + 8C + D = -68 \qquad (1)$$
$$2(-4) + 8(-6) + D = -68$$
$$-8 - 48 + D = -68$$
$$D = -68 + 56$$
$$D = -12 \qquad (8)$$

The solution of the system of equations gave values for three independent constants in the general equation

$$x^2 + y^2 + Bx + Cy + D = 0$$

When the constant values are substituted the equation takes the form of

$$x^2 + y^2 - 4x - 6y - 12 = 0$$

Rearranging and completing the square in x and y,

$$(x^2 - 4x + 4) + (y^2 - 6y + 9) - 12 = 4 + 9$$

$$(x - 2)^2 + (y - 3)^2 = 25$$

which corresponds to a circle with the center at (2, 3) with a radius of 5. This is the circle described by the three given conditions and is shown in figure 10-7 (A).

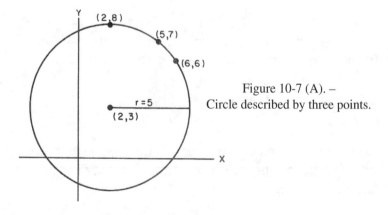

Figure 10-7 (A). – Circle described by three points.

The previous example problem showed one method for determining the equation of a circle when three points are given. The next example shows another method for solving the same problem. One of the most important things to keep in mind when studying analytic geometry is that many problems may be solved by more than one method. Each problem should be analyzed carefully to determine what relationships exist between the given data and the desired results of the problem. Relationships such as distance from one point to another, distance from a point to a line, slope of a line, the Pythagorean theorem, etc., will be used to solve various problems.

EXAMPLE: Find the equation of the circle described by the three points (2, 8), (5, 7), and (6, 6). Use a method other than that used in the previous example problem.

SOLUTION: A different method of solving this problem results from the reasoning in the following paragraphs.

The center of the desired circle will be the intersection of the perpendicular bisectors of the chords connecting points (2, 8) with (5, 7) and (5, 7) with (6, 6), as shown in figure 10-7(B).

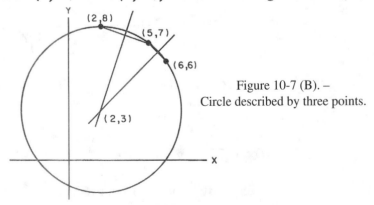

Figure 10-7 (B). – Circle described by three points.

The perpendicular bisector of the line connecting two points is the locus of a point which moves so that it is always equidistant from the two points. Using this analysis we can get the equations of the perpendicular bisectors of the two lines.

Equating the distance formulas which describe the distances from a point (x,y), which is equidistant from the points $(2,8)$ and $(5,7)$, gives

$$\sqrt{(x-2)^2 + (y-8)^2} = \sqrt{(x-5)^2 + (y-7)^2}$$

Squaring both sides gives

$$(x-2)^2 + (y-8)^2 = (x-5)^2 + (y-7)^2$$

or
$$x^2 - 4x + 4 + y^2 - 16y + 64 =$$
$$x^2 - 10x + 25 + y^2 - 14y^2 + 49$$

Canceling and combining terms results in

$$6x - 2y = 6$$

or
$$3x - y = 3$$

Follow the same procedure for the points $(5, 7)$ and $(6, 6)$.

$$\sqrt{(x-5)^2 + (y-7)^2} = \sqrt{(x-6)^2 + (y-6)^2}$$

Squaring each side gives

$$(x-5)^2 + (y-7)^2 = (x-6)^2 + (y-6)^2$$

$$x^2 - 10x + 25 + y^2 + 14y + 49 =$$

$$x^2 - 12x + 36 + y^2 - 12y + 36$$

Canceling and combining terms gives a second equation in x and y.

$$2x - 2y = -2$$

or

$$x - y = -1$$

Solving the equations simultaneously will give the coordinates of the intersection of the two perpendicular bisectors; this is the center of the circle.

$$3x - y = 3$$
$$x - y = -1 \text{ (Subtract)}$$

$$\overline{}$$

$$2x = 4$$
$$x = 2$$

Substitute the value x = 2 in one of the equations to find the value of y.

$$x - y = -1$$
$$2 - y = -1$$
$$-y = -3$$
$$y = 3$$

Thus, the center of the circle is the point (2,3).

The radius will be the distance between the center (2,3) and one of the three given points. Using point (2, 8) we obtain.

$$r = \sqrt{(2 - 2)^2 + (8 - 3)^2} = \sqrt{25} = 5$$

The equation of this circle is

$$(x - 2)^2 + (y - 3)^2 = 25$$

as was found in the previous example.

If a circle is to be defined by three points the points must be noncollinear. In some cases it is obvious that the three points are non-collinear. Such is the case with points (1,1), (-2, 2), and (-1, -1), since the points are in quadrants 1, 2, and 3 respectively and cannot be connected by a straight line. However, there are many cases in which it is difficult to determine by inspection whether or not the points are collinear, and a method for determining this analytically is needed. In the followng example an attempt is made to find the circle described by three points, when the three points are collinear

EXAMPLE: Find the equation of the circle which passes through the points (1,1), (2, 2), (3,3).

SOLUTION: Substitute the given values of x and y in the standard form of the equation of a circle to get three equations in three unknowns.

$$x^2 + y^2 + Bx + Cy + D = 0$$

For (1, 1) $\qquad 1 + 1 + B + C + D = 0$

$$B + C + D = -2 \qquad (9)$$

For (2, 2) $\qquad 4 + 4 + 2B + 2C + D = 0$

$$2B + 2C + D = -8 \qquad (10)$$

For (3, 3) $\qquad 9 + 9 + 3B + 3C + D = 0$

$$3B + 3C + D = -18 \qquad (11)$$

To eliminate D, first subtract (9) from (10).

$$2B + 2C + D = -8$$

$$\underline{B + C + D = -2 \text{ (Subtract)}}$$

$$B + C = -6 \qquad\qquad (12)$$

Next subtract (10) from (11).

$$3B + 3C + D = -18$$
$$\underline{2B + 2C + D = \ \ -8}$$
$$B + C = -10 \qquad (13)$$

Then subtract (13) from (12) to eliminate one of the unknowns.

$$B + C = -6$$
$$\underline{B + C = -10}$$
$$0 + 0 = 4$$
$$0 = 4$$

This solution is not valid and there is no circle through the three given points. The reader should attempt to solve (12) and (13) by the substitution method. When the three given points are collinear an inconsistent solution of some type will result.

If we attempt to solve the problem by eliminating both B and C at the same time (to find D) another type of inconsistent solution results. With the given coefficients it is not difficult to eliminate both A and B at the same time. First, multiply (10) by 3 and (11) by -2 and add the resultant equations.

$$6B + 6C + 3D = -24$$

$$\underline{-6B - 6C - 2D = \ \ 36} \ \ (+)$$

$$D = \ \ 12$$

Then multiply (9) by -2 and add the resultant to (10)

$$-2B - 2C - 2D = 4$$
$$2B + 2C + D = -8$$
$$\overline{-D = -4}$$
$$D = 4$$

This gives two values for D and is inconsistent since each of the constants must have a unique value consistent with the given conditions. The three points are on the straight line $y = x$.

Practice Problems

In each of the problems below find the equation of the circle which passes through the three given points.

1. (14,0), (12,4), and (3,7)
2. (10,3), (11,8), and (7,14)
3. (1,1), (0,0), and (-1,-1)
4. (12,-5), (-9,-12), and (-4,3)

Answers

1. $x^2 + y^2 - 10x + 4y = 56$

2. $x^2 + y^2 - 6x - 14y = 7$

3. No solution; the given points describe the straight line $y = x$.

4. $x^2 + y^2 - 2x + 14y = 75$

The Parabola

The parabola is the locus of all points which are equidistant from a fixed point, called the focus, and a fixed line called the directrix. In the parabola shown in figure 10-8, the point V, which lies halfway between the focus of the directrix, is called the VERTEX of the parabola. In this figure and in many of the parabolas discussed in the first portion of this section, the vertex of the parabola will fall at the origin; however, the vertex of the parabola, like the center of the circle, can fall at any point in the plane.

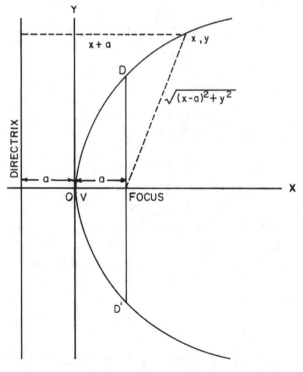

Figure 10-8. – The parabola.

In figure 10-8, the distance from the point (x, y) on the curve to the focus (a, 0) is

$$\sqrt{(x - a)^2 + y^2}$$

The distance from the point (x, y) to the directrix is

$$x + a$$

Since by definition these two distances are equal we may set them equal

$$\sqrt{(x - a)^2 + y^2} = x + a$$

Squaring both sides

$$(x - a)^2 + y^2 = (x + a)^2$$

Expanding

$$x^2 - 2ax + a^2 + y^2 = x^2 + 2ax + a^2$$

Canceling and combining terms we have an equation for the parabola

$$y^2 = 4ax$$

For every positive value of x in the equation of the parabola there are two values of y. But when x becomes negative the values of y are imaginary. Thus, the curve must be entirely to the right of the Y axis when the equation is in this form and a is positive. If the equation is

$$y^2 = - 4\ ax$$

(a negative) the curve lies entirely to the left of the Y axis.

If the form of the equation is

$$x^2 = 4ay$$

the curve will open upward and the focus will be a point on the Y axis. For every positive value of y there will be two values of x and the curve will be entirely above the X axis. When the equation is in the form

$$x^2 = -4ay$$

the curve will open downward, be entirely below the X axis, and have as its focus a point on the negative Y axis. Parabolas which are representative of the four cases given here are shown in figure 10-9.

When x is equal to a in the equation

$$y^2 = 4ax$$

it follows that $\qquad y^2 = 4a^2$

and $\qquad y = \pm 2a$

This value of y is the height of the curve at the focus or the distance from the focus to point D in figure 10-8. The width of the curve at the focus is the distance from point D to point D' in the figure and is equal to 4a. This width is called the LATUS RECTUM in many texts; however, a more descriptive term is FOCAL CHORD and both terms will be used in this course. The latus rectum is one of the properties of a parabola which is used in the analysis of a parabola or in the sketching of a parabola.

EXAMPLE: Give the value of a, the length of the focal chord, and the equation of the parabola which is the locus of all points equidistant from the point (3, 0) and the line x = -3.

SOLUTION: First plot the given information on a coordinate system as shown in figure 10-10 (A). Reference to figure 10-8 shows that the point (3, 0) corresponds to the position of the focus and that the line x = -3 is the directrix of the parabola. Figure 10-8 also shows that the value of a is equal to one half of the distance from the focus to the directrix or, in this problem, one half the distance from x = -3 to x = 3. Thus, the value of a is 3.

The second value required by the problem is the length of the focal chord. As stated previously, the focal chord length is equal to 4a. The value of a was found to be 3 so the length of the focal chord is 12. Reference to figure 10-8 shows that one extremity of the focal chord will be a point on the curve which is 2a or 6 units above the focus, and the other extremity is a second point 2a or 6 units below the focus. Using this information and recalling that the vertex is one-half the distance from the focus to the directrix, plot three more points as shown in figure 10-10 (B).

Now a smooth curve through the vertex and the two points that are the extremities of the focal chord is a sketch of the parabola in this problem. (See fig. 10-10 (C).)

To find the equation of the hyperbola refer to figure 10-10 (D) and use the procedure used earlier. We know by definition that any point P(x, y) on the parabola is equidistant from the focus and directrix.

Thus we equate these two distances and

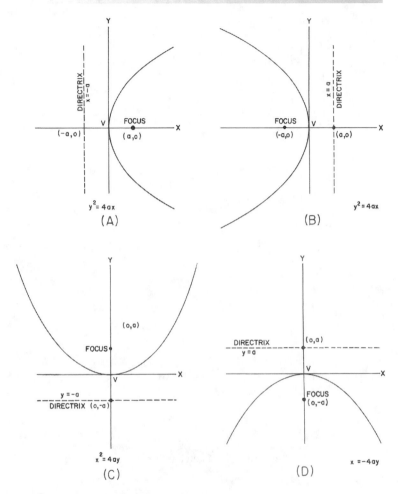

Figure 10-9. – Parabolas corresponding to four forms of the equation.

$$\sqrt{(x - a)^2 + y^2} = x + a$$

However, we have found the distance a to be equal to 3 so we substitute and

$$(x - 3)^2 + y^2 = x + 3$$

113

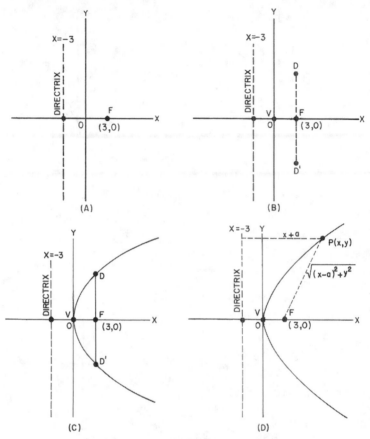

Figure 10-3. – Sketch of a parabola.

Square both sides $(x - 3)^2 + y^2 = (x + 3)^2$

Expand $x^2 - 6x + 9 + y^2 = x^2 + 6x + 9$

Cancel and combine terms to obtain the equation of the parabola

$$y^2 = 12x$$

If we check the consistency of our findings, we see that the form of the equation and the

114

sketch agree with figure 10-9 (A). Also, the 12 in the right side of the equation corresponds to the 4a in the general form and is correct since we determined that the value of a was 3.

NOTE: When the focus of a parabola lies on the Y axis, the equated distance equation is

$$\sqrt{(y - a)^2 + x^2} = y + a$$

Practice Problems

Give the equation, the value of a, and the length of the focal chord of the parabola which is the locus of all points equidistant from the point and line given in the following problems.

1. The point (-2,0) and the line x = 2
2. The point (0,4) and the line y = -4
3. The point (0,-1) and the line y = 1
4. The point (1,0) and the line x = -1

Answers

1. $y^2 = -8x$, a = -2, f.c. = 8
2. $x^2 = 16y$, a = 4, f.c. = 16
3. $x^2 = -4y$, a = -1, f.c. = 4
4. $y^2 = 4x$, a = 1, f.c. = 4

Formula Generalization

All of the parabolas in the preceding section had the vertex at the origin and the corresponding equations were in one of four forms as follows:

115

1. $y^2 = 4ax$ 3. $x^2 = 4ay$

2. $y^2 = -4ax$ 4. $x^2 = -4ay$

In this section we will present four more forms of the equation of a parabola, generalized to consider a parabola with a vertex at point V (h,k). When the vertex is moved from the origin to a point V(h,k) the x and y terms of the equation are replaced by (x - h) and (y - k). Then the general equation for the parabola which opens to the right (fig. 10-9 (A)) is

$$(y - k)^2 = 4a(x - h)$$

The four general forms of the equations for parabolas with vertex at the point V(h,k) are as follows:

1. $(y - k)^2 = 4a (x - h)$, corresponding to $y^2 = 4ax$, parabola opens to the right

2. $(y - k)^2 = -4a (x - h)$, corresponding to $y^2 = -4ax$, parabola opens to the left

3. $(x - h)^2 = 4a (y - k)$, corresponding to $x^2 = 4ay$, parabola opens upward

4. $(x - h)^2 = -4a (y - k)$, corresponding to $x^2 = -4ay$, parabola opens downward.

The method for reducing an equation to one of these standard forms is similar to the methods used for reducing the equation of a circle.

EXAMPLE: Reduce the equation

$$y^2 - 6y - 8x + 1 = 0$$

to standard form.

SOLUTION: Rearrange the equation so that the second degree term and any first degree terms of the same unknown are on the left side. Then group the unknown term which appears only in the first degree and all constants on the right.

$$y^2 - 6y = 8x - 1$$

Then complete the square in y

$$y^2 - 6y + 9 = 8x - 1 + 9$$

$$(y - 3)^2 = 8x + 8$$

To get the equation in the form

$$(y - k)^2 = 4a(x - h)$$

factor an 8 out of the right side. Thus

$$(y - 3)^2 = 8(x + 1)$$

is the equation of the parabola.

Practice Problems

Reduce the equations given in the following problems to standard form.

1. $x^2 + 4 = 4y$

2. $y^2 - 4x = 6y + 9$

3. $4x + 8y + y^2 + 20 = 0$

4. $4x - 12y + 40 + x^2 = 0$

Answers

1. $x^2 = 4(y - 1)$ 3. $(y + 4)^2 = -4(x + 1)$

2. $(y - 3)^2 = 4x$ 4. $(x + 2)^2 = 12(y - 3)$

The Ellipse

An ellipse is a conic section with an eccentricity less than one.

Referring to figure 10-11, let

$$PO = a$$
$$FO = c$$
$$OM = d$$

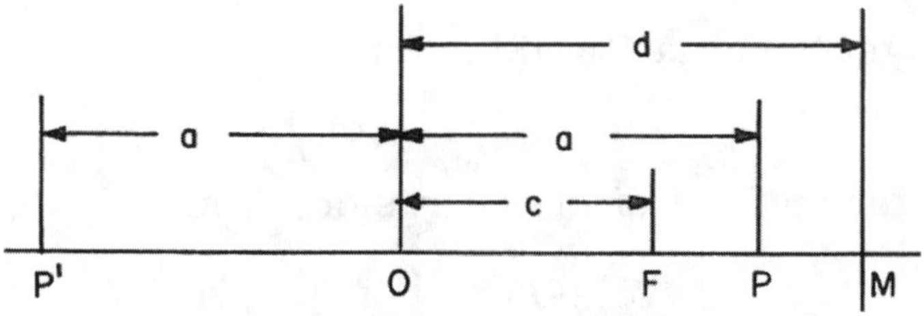

Figure 10-11. – Development of focus and directrix.

where F is the focus, 0 is the center, and P and P' are points on the ellipse. Then from the definition of eccentricity,

$$\frac{a - c}{d - a} = e \quad a - c = ed - ea$$

$$\frac{a + c}{d + a} = e \quad a + c = ed + ea$$

Addition and subtraction of the two equations give:

$$2c = 2ae \text{ or } c = ae$$
$$2a = 2de \text{ or } d = \frac{a}{e} \tag{14}$$

118

Place the center of the ellipse at the origin so that one focus lies at $(-ae, 0)$ and one directrix is the line $x = -a/e$.

Referring to figure 10-12, there will be a point on the Y axis which will satisfy the conditions for an ellipse. Let

$$P''O = b \qquad\qquad FO = c$$

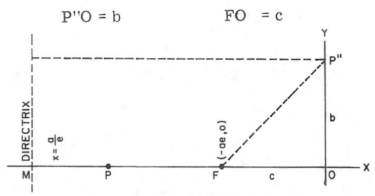

Figure 10-12. – Focus directrix, and point P".

Then
$$P''F = \sqrt{b^2 + c^2}$$

and the ratio of the distance of P" from the focus and the directrix is e so that

$$\frac{\sqrt{b^2 + c^2}}{\dfrac{a}{e}} = e$$

Multiplying both sides by a/e gives

$$\sqrt{b^2 + c^2} = a \quad\text{or}\quad b^2 + c^2 = a^2$$

so that
$$b = \pm\sqrt{a^2 - c^2} \qquad (15)$$

Now combining equations (14) and (15) gives

$$b = \pm\sqrt{a^2 - a^2 e^2} \quad\text{or}\quad b = \pm a\sqrt{1 - e^2} \qquad (16)$$

119

Refer to figure 10-13. If the point (x, y) is on the ellipse, the ratio of its distance from F to its distance from the directrix will be e: The distance from (x, y) to the focus (-ae,0) will be

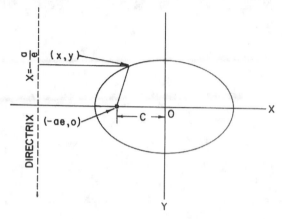

Figure 10-13. – The ellipse.

$$\sqrt{(x + ae)^2 + y^2}$$

and the distance from (x, y) to the directrix

$x = -\dfrac{a}{e}$ is $x + \dfrac{a}{e}$

The ratio of these two distances is equal to e so that

$$\frac{\sqrt{(x + ae)^2 + y^2}}{x + \dfrac{a}{e}} = e$$

or $\quad \sqrt{(x + ae)^2 + y^2} = e\left(x + \dfrac{a}{e}\right)$

$$= ex + a$$

Squaring both sides gives

$$x^2 + 2aex + a^2e^2 + y^2 = e^2x^2 + 2aex + a^2$$

Canceling like terms and transposing terms in x to the left-hand side of the equation gives

$$x^2 - e^2x^2 + y^2 = a^2 - a^2e^2$$

Removing a common factor,

$$x^2(1 - e^2) + y^2 = a^2(1 - e^2) \qquad (17)$$

Dividing equation (17) through by the right-hand member,

$$\frac{x^2}{a^2} + \frac{y^2}{a^2(1 - e^2)} = 1$$

From equation (16) we obtain

$$b = \pm a\sqrt{1 - e^2}$$

$$a^2(1 - e^2) = b^2$$

so that the equation becomes

$$\frac{x^2}{a^2} + \frac{y^2}{b^2} = 1 \qquad (18)$$

This is the equation of an ellipse in standard form. In figure 10-14, a is the length of the semimajor axis and b is the length of the semiminor axis.

The curve is symmetrical with respect to the x and y axes, so that it is easily seen that it has another focus at (ae,0) and a corresponding directrix x = a/e.

The distance from the center through the focus to the curve is always designated a and is called the semimajor axis. This axis may be in either the x or y direction. When it is in the y direction, the directrix is a line with the equation y = k

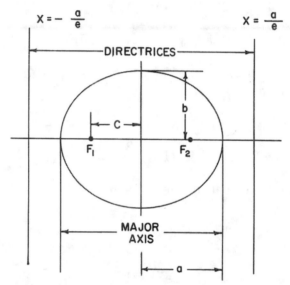

Figure 10-14. – Ellipse showing axes.

In the case we have studied, the directrix was denoted by the formula

$$x = k$$

where k is a constant equal to - a/e.

The perpendicular distance from the midpoint of the major axis to the curve is called the semiminor axis and is always signified by b.

The distance from the center of the ellipse to the focus is called c and in any ellipse the following relations hold for a, b, and c

$$c = \sqrt{a^2 - b^2}$$

$$b = \sqrt{a^2 - c^2}$$

$$a = \sqrt{b^2 + c^2}$$

Whenever the directrix is a line with the equation $y = k$ the major axis will be in the y direction and the equation of the ellipse will be as follows:

$$\frac{x^2}{b^2} + \frac{y^2}{a^2} = 1 \tag{19}$$

Otherwise everything remains as before and the equation is given by (18).

In an ellipse the position of the a^2 and b^2 terms indicate the orientation of the ellipse axis. As shown in figure 10-14 the value is the semimajor or longer axis.

In the previous paragraphs formulas were given which related a, b, and c and, in the first portion of this discussion, a formula relating a, c, and the eccentricity was given. These relationships will be used to find the equation of an ellipse in the following example.

EXAMPLE: Find the equation of the ellipse with center at the origin and having foci at $(\pm 2\sqrt{6}, 0)$ and an eccentricity equal to $\frac{2\sqrt{6}}{7}$.

SOLUTION: With the focal points on the x axis the ellipse is oriented as in figure 10-14 and the standard form of the equation is

$$\frac{x^2}{a^2} + \frac{y^2}{b^2} = 1$$

123

With the center at origin the numerators of the fractions on the left are x^2 and y^2 so the problem is to find the values of a and b.

The distance from the center to either of the foci is the value c (fig. 10-14) so in this problem

$$c = \pm 2\sqrt{6}$$

from the given coordinates of the foci.

The values of a, c, and e (eccentricity) are related by

$$c = ae \quad \text{or} \quad a = \frac{c}{e}$$

From the known information, substitute the values of c and e

$$a = \frac{\pm 2\sqrt{6}}{\dfrac{2\sqrt{6}}{7}}$$

$$a = \pm 2\sqrt{6} \times \frac{7}{2\sqrt{6}}$$

and

$$a = \pm 7$$

$$a^2 = 49$$

Then, using the formula

$$b = \sqrt{a^2 - c^2} \quad \text{or} \quad b^2 = a^2 - c^2$$

and substituting for a^2 and c^2

$$b^2 = 49 - (\pm 2\sqrt{6})2$$
$$b^2 = 49 - (4 \times 6)$$
$$b^2 = 49 - 24$$

gives the final required value of

$$b^2 = 25$$

Then, the equation of the ellipse is

$$\frac{x^2}{49} + \frac{y^2}{25} = 1$$

Practice Problems

Find the equation of the ellipse with center at the origin and for which the following properties are given.

1. Foci at $(\pm \sqrt{7}, 0)$ and an eccentricity of $\frac{\sqrt{7}}{4}$

2. $b = 5$, $e = \frac{\sqrt{11}}{6}$

3. $a = 7$, $e = \frac{3\sqrt{5}}{7}$

Answers

1. $\dfrac{x^2}{4^2} + \dfrac{y^2}{3^2} = 1$

2. $\dfrac{x^2}{36} + \dfrac{y^2}{25} = 1$

3. $\dfrac{x^2}{7^2} + \dfrac{y^2}{2^2} = 1$

Ellipse as a Locus of Points

An ellipse may be defined as the locus of a point which moves so that the sum of its distances from two fixed points is a constant equal to 2a.

Let the foci be F_1 and F_2 at $(\pm ae, 0)$, as shown in figure 10-15, and let the directrices be

$$x = \frac{\pm a}{e}$$

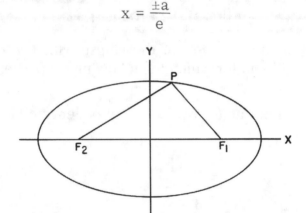

Figure 10-15. – Ellipse, center at origin.

Then
$$F_1 P = e \left(\frac{a}{e} - x \right) = a - ex$$

$$F_2 P = e \left(\frac{a}{e} + x \right) = a + ex$$

so that $F_1 P + F_2 P = a - ex + a + ex$

$$F_1 P + F_2 P = 2a$$

Whenever the center of the ellipse is at some point other than (0,0), say at the point (h,k), figure 10-16, the equation of the ellipse must be modified to the following form

126

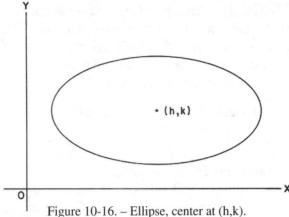

Figure 10-16. – Ellipse, center at (h,k).

$$\frac{(x - h)^2}{a^2} + \frac{(y - k)^2}{b^2} = 1 \qquad (20)$$

Subtracting h from the value of x reduces the value of the term (x - h) to the value which x would have if the center were at the origin. The term (y - k) is identical in value to the value of y if the center were at the origin.

Reduction to a Standard Form

Whenever we have an equation in the form

$$Ax^2 + Cy^2 + Dx + Ey + F = 0 \qquad (21)$$

where the capital letters refer to independent constants and A and C have the same sign, we can reduce the equation to the standard form for an ellipse. Completing the squares in x and y and performing a few simple algebraic transformations will change the form to that of equation (20).

127

THEOREM: An equation of the second degree, in which the xy term is missing and the coefficients of x^2 and y^2 are different but have the same sign, represents an ellipse with axes parallel to the coordinate axes.

EXAMPLE: Reduce the equation

$$4x^2 + 9y^2 - 40x - 54y + 145 = 0$$

to the standard form of an ellipse.

SOLUTION: Collect terms in x and y and remove the common factors of these terms.

$$4x^2 - 40x + 9y^2 - 54y + 145 = 0$$

$$4(x^2 - 10x) + 9(y^2 - 6y) + 145 = 0$$

Transpose the constant terms and complete the squares in x and y. When there are factored terms involved in completing the square, as in this example, an error is frequently made. The factored value operates on the term added inside the parentheses as well as the original terms. Therefore, the values added to the right side of the equation will be the product of the factored value and the term added to complete the square.

$$4(x^2 - 10x + 25) + 9(y^2 - 6y + 9)$$

$$= -145 + 4(25) + 9(9)$$

$$= -145 + 100 + 81$$

$$= 36$$

$$4(x - 5)^2 + 9(y - 3)^2 = 36$$

Divide through by the right-hand (constant) term. This reduces the right member to 1 as required by the standard form.

$$\frac{4(x - 5)^2}{36} + \frac{9(y - 3)^2}{36} = 1$$

$$\frac{(x - 5)^2}{9} + \frac{(y - 3)^2}{4} = 1$$

This reduces to the standard form

$$\frac{(x - 5)^2}{(3)^2} + \frac{(y - 3)^2}{(2)^2} = 1$$

Corresponding to equation (20) and represents an ellipse with the center at (5,3), its semimajor axis (a) equal to 3, and its semiminor axis (b) equal to 2.

EXAMPLE: Reduce the equation

$$3x^2 + y^2 + 20x + 32 = 0$$

to the standard form of an ellipse.

SOLUTION: First, collect terms in x and y. As in the previous example, the coefficients of x^2 and y^2 must be reduced to 1 in order to facilitate completing the square. Thus the coefficient of the x^2 term is divided out of the two terms containing x, as follows:

$$3x^2 + 20x + y^2 + 32 = 0$$

$$3\left(x^2 + \frac{20x}{3}\right) + y^2 = -32$$

Complete the square in x noting that there will be a product added to the right side

$$3\left(x^2 + \frac{20x}{3} + \frac{100}{9}\right) + y^2 = -32 + 3\left(\frac{100}{9}\right)$$

129

$$3\left(x + \frac{10}{3}\right)^2 + y^2 = -32 + \frac{300}{9}$$

$$3\left(x + \frac{10}{3}\right)^2 + y^2 = \frac{-288 + 300}{9}$$

$$3\left(x + \frac{10}{3}\right)^2 + y^2 = \frac{12}{9}$$

$$3\left(x + \frac{10}{3}\right)^2 + y^2 = \frac{4}{3}$$

Divide through by the right-hand term.

$$3\frac{\left(x + \frac{10}{3}\right)^2}{\frac{4}{3}} + \frac{y^2}{\frac{4}{3}} = 1$$

$$9\frac{\left(x + \frac{10}{3}\right)^2}{4} + \frac{y^2}{\frac{4}{3}} = 1$$

$$\frac{\left(x + \frac{10}{3}\right)^2}{\left(\frac{4}{9}\right)} + \frac{y^2}{\left(\frac{4}{3}\right)} = 1$$

This reduces to the standard form

$$\frac{\left(x + \frac{10}{3}\right)^2}{\left(\frac{2}{3}\right)^2} + \frac{y^2}{\left(\frac{2}{\sqrt{3}}\right)^2} = 1$$

130

$$\frac{\left(x + \frac{10}{3}\right)^2}{\left(\frac{2}{3}\right)^2} + \frac{y^2}{\left(\frac{2\sqrt{3}}{3}\right)^2} = 1$$

Practice Problems

Express the following equations in the standard form for an ellipse.

1. $5x^2 - 110x + 4y^2 + 425 = 0$

2. $x^2 - 14x + 36y^2 - 216y + 337 = 0$

3. $9x^2 - 54x + 4y^2 + 16y + 61 = 0$

4. $3x^2 - 14x + 4y^2 + 11 = 0$

Answers

1. $\dfrac{(x - 11)^2}{(6)^2} + \dfrac{y^2}{(3\sqrt{5})^2} = 1$

2. $\dfrac{(x - 7)^2}{(6)^2} + \dfrac{(y - 3)^2}{1} = 1$

3. $\dfrac{(x - 3)^2}{(2)^2} + \dfrac{(y + 2)^2}{(3)^2} = 1$

4. $\dfrac{\left(x + \frac{7}{3}\right)^2}{\left(\frac{4}{3}\right)^2} + \dfrac{y^2}{\left(\frac{2\sqrt{3}}{3}\right)^2} = 1$

131

The Hyperbola

A hyperbola is a conic section with an eccentricity greater than one.

The formulas

$$c = ae \quad \text{and} \quad d = \frac{a}{e}$$

developed in the section concerning the ellipse were derived so that they hold true for any value of eccentricity. Thus, they hold true for the hyperbola as well as for an ellipse. Since e is greater than one for a hyperbola, then

$$c = ae \text{ and } c > a$$
$$d = \frac{a}{e} \text{ and } d < a$$

Therefore $c > a > d$.

According to this analysis, if the center of symmetry of a hyperbola is the origin, the foci will lie farther from the origin than the directrices. An inspection of figure 10-17 shows that the curve will never cross the Y axis. Thus, the solution for the value of b, the semiminor axis of the ellipse will yield no real value for b. In other words, b will be an imaginery number. This can easily be seen from the equation

$$b = \sqrt{a^2 - c^2}$$

since $c > a$ for a hyperbola.

However, we can square both sides of the above equation, and since the square of an imaginary number is a negative real number we write

$$-b^2 = a^2 - c^2 \quad \text{or} \quad b^2 = c^2 - a^2$$

and, since c = ae,

$$b^2 = a^2e^2 - a^2 = a^2(e^2 - 1)$$

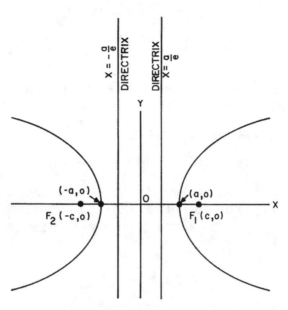

Figure 10-17. – The hyperbola.

Now we can use this equation to obtain the equation of a hyperbola from the following equation which was developed in the section on the ellipse.

$$\frac{x^2}{a^2} + \frac{y^2}{a^2(1 - e^2)} = 1$$

and since $a^2(1 - e^2) = -a^2(e^2 - 1) = -b^2$

we have $\dfrac{x^2}{a^2} - \dfrac{y^2}{b^2} = 1$

133

This is a standard form for the equation of a hyperbola. The solution of this equation for y gives

$$y = \pm\frac{b}{a}\sqrt{x^2 - a^2}$$

which shows that y is imaginary only when $x^2 < a^2$. The curve, therefore, lies entirely beyond the two lines $x = \pm a$ and crosses the x axis at $x = \pm a$.

The two straight lines

$$bx + ay = 0 \text{ and } bx - ay = 0 \qquad (22)$$

can be used to illustrate an interesting property of a hyperbola. The distance from the line $bx - ay = 0$ to a point (x_1, y_1) on the curve is given by

$$d = \frac{bx_1 - ay_1}{\sqrt{a^2 + b^2}} \qquad (23)$$

Since (x_1, y_1) is on the curve, its coordinates satisfy the equation

$$b^2 x_1^2 - a^2 y_1^2 = a^2 b^2$$

which may be written

$$(bx_1 - ay_1)(bx_1 + ay_1) = a^2 b^2$$

or $\qquad bx_1 - ay_1 = \dfrac{a^2 b^2}{bx_1 + ay_1}$

Now substituting this value into equation (23), gives us

$$d = \frac{a^2 b^2}{\sqrt{a^2 + b^2}} \left(\frac{1}{bx_1 + ay_1} \right)$$

As the point (x_1, y_1) is chosen farther and farther from the center of the hyperbola, the absolute values for x_1 and y_1 will increase and the distance d will approach zero. A similar result can easily be derived for the line $bx + ay = 0$.

The lines of equation (22) which are usually written

$$y = -\frac{b}{a} x \text{ and } y = +\frac{b}{a}x$$

are called the asymptotes of the hyperbola. They are very important in tracing a curve and studying its properties. The asymptotes of a hyperbola, figure 10-18, are the diagonals of the rectangle whose center is the center of the curve and whose sides are parallel and equal to the axes of the curve. The latus rectum of a hyperbola is equal to $\frac{2b^2}{|a|}$

Another definition of a hyperbola is the locus of a point that moves so that the difference of its distances from two fixed points is constant. The fixed points are the foci and the constant difference is 2a.

The nomenclature of the hyperbola is slightly different from that of an ellipse. The transverse axis is 2a or the distance between the intersections of the hyperbola with its focal axis.

135

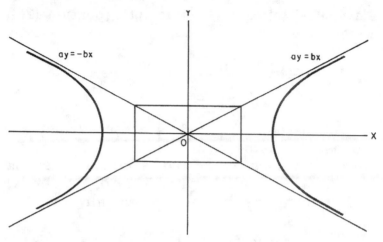

Figure 10-18. – Using asymptotes to sketch a hyperbola.

The conjugate axis is 2b and is perpendicular to the transverse axis.

Whenever the foci are on the Y axis and the directrices are lines of the form $y = \pm k$, where k is a constant, the equation of the hyperbola will read

$$\frac{y^2}{a^2} - \frac{x^2}{b^2} = 1$$

This equation represents a hyperbola with its transverse axis on the Y axis. Its asymptotes are the lines $by - ax = 0$ and $by + ax = 0$.

Analysis of the Equation

The properties of the hyperbola most often used in analysis of the curve are the foci, directrices, length of the latus rectum, and the equations of the asymptotes.

Reference to figure 10-17 shows that the foci are given by the points F_1 $(c,0)$ and F_2 $(-c,0)$ when the equation of the hyperbola is in the form

$$\frac{x^2}{a^2} - \frac{y^2}{b^2} = 1$$

If the equation were

$$\frac{y^2}{b^2} - \frac{x^2}{a^2} = 1$$

the foci would be the points $(0,c)$ and $(0,-c)$. The value of c is either determined from the formula

$$c^2 = a^2 + b^2$$

or the formula $c = ae$

Figure 10-17 also shows that the directrices are the lines $x = \pm\frac{a}{e}$ or, in the case where the hyperbolas open upward and downward, $y = \pm\frac{a}{e}$. This is also given earlier in this discussion as $d = \frac{a}{e}$.

The equations of the asymptotes are given earlier as

$$bx + ay = 0 \text{ and } bx - ay = 0$$

or $\qquad y = -\frac{b}{a}x \text{ and } y = +\frac{b}{a}x$

It was also pointed out that the length of the latus rectum is equal to $\frac{2b^2}{|a|}$

The properties of a hyperbola can be determined from the equation of a hyperbola or the equation can be written given certain properties, as shown in the following examples. In these examples and in the practice problems immediately following, all of the hyperbolas considered have their centers at the origin.

EXAMPLE: Find the equation of the hyperbola with an eccentricity of 3/2, directrices x = ±4/3, and foci at (±3,0).

SOLUTION: The foci lie on the X axis at the points (3,0) and (-3,0) so the equation is of the form

$$\frac{x^2}{a^2} - \frac{y^2}{b^2} = 1$$

This fact is also shown by the equation of the directrices.

Before proceeding with the problem one point should be emphasized: in the basic formula for the hyperbola the a^2 term will always be the denominator for the x^2 term and the b^2 term the denominator for the y^2 term. The orientation of the axis of symmetry is not dependent on the size of a^2 and b^2 as in the ellipse; it lies along or parallel to the axis of the positive x^2 or y^2 term. Since we have determined the form of the equation and since the center of the curve in this section is restricted to the origin the problem is reduced to finding the values of a^2 and b^2.

First, the foci are given as (±3,0) and since the foci are also the points (±c,0) it follows that

$$c = \pm 3$$

The eccentricity is given and the value of a^2 can be determined from the formula

$$c = ae$$

$$a = \frac{c}{e}$$

$$a = \frac{\pm 3}{\frac{3}{2}}$$

$$a = \frac{\pm 6}{3}$$

$$a = \pm 2$$

$$a^2 = 4$$

The relationship of a, b, and c for the hyperbola is

$$b^2 = c^2 - a^2$$

and

$$b^2 = (\pm 3)^2 - (\pm 2)^2$$

$$b^2 = 9 - 4$$

$$b^2 = 5$$

When these values are substituted in the equation

$$\frac{x^2}{a^2} - \frac{y^2}{b^2} = 1$$

the equation

$$\frac{x^2}{4} - \frac{y^2}{5} = 1$$

results and is the equation of the hyperbola.

The equation could also be found by the use of other relationships which utilize the given information.

The directrices are given as

$$x = \pm \frac{4}{3}$$

and, since

$$d = \frac{a}{e}$$

or

$$a = de$$

Substituting the values given for d and e results in

$$a = \pm \frac{4}{-3} \left(\frac{3}{2} \right)$$

When $d > 0$

$$a = \frac{4}{3} \left(\frac{3}{2} \right)$$

$$a = 2$$

When $d < 0$

$$a = -\frac{4}{3} \left(\frac{3}{2} \right)$$

$$a = -2$$

therefore

$$a = \pm 2$$

and

$$a^2 = 4$$

While the value of c can be determined by the given information in this problem, it could also be computed since

$$c = ae$$

and a has been found to equal ± 2 and e is given as $\frac{3}{2}$, then

$$c = \pm 2 \left(\frac{3}{2} \right)$$

and, when $a > 0$

$$c = 2 \left(\frac{3}{2}\right)$$

$$c = 3$$

For $a < 0$

$$c = -2 \left(\frac{3}{2}\right)$$

$$c = -3$$

Then

$$c = \pm 3$$

With values for a and c computed, the value of b is found as before and the equation can be written.

EXAMPLE: Find the foci, directrices, eccentricity, length of the latus rectum, and equations of the asymptotes of the hyperbola described by the equation

$$\frac{x^2}{9} - \frac{y^2}{16} = 1$$

SOLUTION: This equation is of the form

$$\frac{x^2}{a^2} - \frac{y^2}{b^2} = 1$$

and the values for a and b are determined by inspection to be

$$a^2 = 9$$

$$a = \pm 3$$

and

$$b^2 = 16$$

$$b = \pm 4$$

With a and b known, find c by using the formula

$$b^2 = c^2 - a^2$$

$$c^2 = a^2 + b^2$$

$$c = \pm \sqrt{a^2 + b^2}$$

$$c = \pm \sqrt{9 + 16}$$

$$c = \pm \sqrt{25}$$

$$c = \pm 5$$

From the form of the equation we know that the foci are at the points

$$F_1(c, 0)$$

and

$$F_2(-c, 0)$$

so the foci = $(\pm 5, 0)$.

The eccentricity is found by the formula

$$e = \frac{c}{a}$$

$$e = \frac{\pm 5}{\pm 3}$$

$$e = \frac{5}{3}$$

Reference to figure 10-17 shows that with the center at the origin, c and a will have the same sign.

The directrix is found by the formula

$$d = \frac{a}{e}$$

or, since this equation will have directrices parallel to the Y axis, use the formula

$$x = \frac{a}{e}$$

Then

$$x = \frac{\pm 3}{\frac{5}{3}}$$

$$x = \pm 3 \left(\frac{3}{5}\right)$$

When a > 0

$$x = 3 \left(\frac{3}{5}\right)$$

$$x = \frac{9}{5}$$

and when a < 0

$$x = -3 \left(\frac{3}{5}\right)$$

$$x = -\frac{9}{5}$$

so the directrices are the lines

$$x = \frac{\pm 9}{5}$$

The latus rectum (l. r.) is found by

$$l.\ r.\ = \frac{2b^2}{|a|}$$

143

$$\text{l. r.} = \frac{2(16)}{3}$$

$$\text{l. r.} = \frac{32}{3}$$

Finally, the equations of the asymptotes are the equation of the two straight lines

$$bx + ay = 0$$

and $$bx - ay = 0$$

In this problem, substituting the values of a and d in the equation gives

$$4x + 3y = 0$$

and $$4x - 3y = 0$$

or $$4x \pm 3y = 0$$

The equations of the lines asymptotic to the curve can also be written in the form

$$y = \frac{b}{a}x$$

and $$y = -\frac{b}{a}x$$

In this form the lines are

$$y = \frac{4}{3}x$$

and $$y = -\frac{4}{3}x$$

or $$y = \pm\frac{4}{3}x$$

144

If we think of this equation as a form of the slope intercept formula

$$y = mx + b$$

from chapter 9, the lines would have slopes of $\pm \dfrac{b}{a}$ and each would have its Y intercept at the origin as shown in figure 10-18.

Practice Problems

1. Find the equation of the hyperbola with an eccentricity of $\sqrt{2}$, directrices $x = \pm \dfrac{\sqrt{2}}{2}$, and foci at $(\pm\sqrt{2},\ 0)$.

2. Find the equation of the hyperbola with an eccentricity of $5/3$, foci at $(\pm 5, 0)$, and directrices $x = \pm\ 9/5$.

Find the foci, directrices, eccentricity, equations of the asymptotes, and length of the latus rectum of the hyperbolas given in problems 3 and 4.

3. $\dfrac{x^2}{9} - \dfrac{y^2}{9} = 1$

4. $\dfrac{x^2}{9} - \dfrac{y^2}{4} = 1$

Answers

1. $x^2 - y^2 = 1$

2. $\dfrac{x^2}{9} - \dfrac{y^2}{16} = 1$

3. foci = $(\pm 3\sqrt{2}, 0)$; directrices, x $=\dfrac{\pm 3}{\sqrt{2}}$, eccentricity = $\sqrt{2}$; l.r. = 6; asymptotes $x \pm y = 0$.

4. foci = $(\pm\sqrt{13}, 0)$; directrices x $=\dfrac{\pm 9}{\sqrt{13}}$; eccentricity $=\dfrac{\sqrt{13}}{3}$; l. r. $=\dfrac{8}{3}$; asymptotes $2x \pm 3y = 0$.

The hyperbola can be represented by an equation in the form

$$Ax^2 + Cy^2 + Dx + Ey + F = 0$$

where the capital letters refer to independent constants and A and C have different signs. These equations can be reduced to standard form in the same manner in which similar equations for the ellipse were reduced to standard form. The general forms of these standard equations are given by

$$\frac{(x - h)^2}{a^2} - \frac{(y - k)^2}{b^2} = 1$$

and

$$\frac{(y - k)^2}{b^2} - \frac{(x - h)^2}{a^2} = 1$$

Polar Coordinates

So far we have located a point in a plane by giving the distances of the point from two perpendicular lines. The location of a point can be defined equally well by noting its distance and bearing. This method is commonly used aboard ship to show the position of another ship or

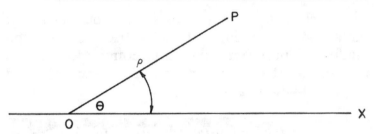

Figure 10-19. – Defining the polar coordinates.

target. Thus, 3 miles at 35° locates the position of a ship relative to the course of the ship making the reading. We can use this method to develop curves and bring out their properties. Assume a fixed direction OX and a fixed point 0 on the line in figure 10-19. The position of any point P is fully determined, if we know the directed distance from 0 to P and the angle that the line OP makes with reference line OX. The line OP is called the radius vector and the angle POX is the polar angle. The radius vector is designed ρ while θ is the angle designation.

Point 0 is the pole or origin. As in conventional trigonometry, the polar angle is positive when measured counterclockwise and negative when measured clockwise. However, unlike the convention established in trigonometry, the radius vector for polar coordinates is positive only when it is laid off on the terminal side of the angle. When the radius vector is laid off on the terminal side of the ray produced beyond the pole (the given angle plus 180°) a negative value is assigned the radius vector. For this reason, there may be more than one equation in polar coordinates to describe a given locus. The concept of a negative radius vector is utilized in some advanced mathematics.

For purposes of this course the concept is not explained or used. It is sufficient that the reader remember that the convention of an always positive radius vector is not followed in some branches of mathematics.

Transformation from Cartesian to Polar Coordinates

At times it will be simpler to work with the equation of a curve in polar coordinates than in cartesian coordinates. Therefore, it is important to know how to change from one system to the other. Sometimes both forms are useful, for some properties of the curve may be more apparent from one form of the equation and other properties more evident from the other.

Transformations are made by applying the following equations which can be derived from figure 10-20.

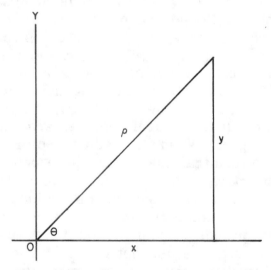

Figure 10-20. – Cartesian and polar relationship.

$$x = \rho \cos \theta \qquad\qquad (24)$$

$$y = \rho \sin \theta \qquad\qquad (25)$$

$$\rho^2 = x^2 + y^2 \qquad\qquad (26)$$

$$\tan \theta = \frac{y}{x} \qquad\qquad (27)$$

EXAMPLE: Change the equation

$$y = x^2$$

from rectangular to polar coordinates.
   SOLUTION: Substitute $\rho \cos \theta$ for x and $\rho \sin \theta$ for y so that we have

$$\rho \sin \theta = \rho^2 \cos^2 \theta$$

$$\sin \theta = \rho \cos^2 \theta$$

or

$$\rho = \frac{\sin \theta}{\cos^2 \theta}$$

$$\rho = \tan \theta \sec \theta$$

EXAMPLE: Express the equation of the circle with its center at (a,0) and with a radius a, as shown in figure 10-21.

$$(x - a)^2 + y^2 = a^2$$

in polar coordinates.

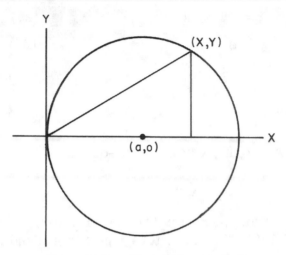

Figure 10-21. – Circle with center (a,0).

SOLUTION: First, expanding this equation gives us

$$x^2 - 2ax + a^2 + y^2 = a^2$$

Rearranging terms we have

$$x^2 + y^2 = 2ax$$

The use of equation (26) gives us

$$\rho^2 = 2ax$$

and applying the value of x given by equation (24), results in

$$\rho^2 = 2a\rho \cos \theta$$

Dividing through by ρ we have the equation of a circle with its center at (a,0) and radius a in polar coordinates

$$\rho = 2a \cos \theta$$

150

Transformation from Polar to Cartesian Coordinates

In order to transform to an equation in cartesian or rectangular coordinates from an equation in polar coordinates use the following equations which can be derived from figure 10-22.

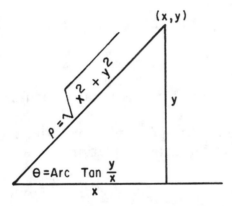

Figure 10-22. – Polar to cartesian relationship.

$$\rho = \sqrt{x^2 + y^2} \tag{28}$$

$$\cos \theta = \frac{x}{\sqrt{x^2 + y^2}} \tag{29}$$

$$\sin \theta = \frac{y}{\sqrt{x^2 + y^2}} \tag{30}$$

$$\tan \theta = \frac{y}{x} \tag{31}$$

$$\sec \theta = \frac{\sqrt{x^2 + y^2}}{x} \tag{32}$$

$$\csc \theta = \frac{\sqrt{x^2 + y^2}}{y} \tag{33}$$

$$\cot \theta = \frac{x}{y} \tag{34}$$

EXAMPLE: Change the equation

$$\rho = \sec \theta \, \tan \theta$$

to an equation in rectangular coordinates

SOLUTION: Applying relations (28), (31), and (32) to the above equation gives

$$\sqrt{x^2 + y^2} = \frac{\sqrt{x^2 + y^2}}{x} \left(\frac{y}{x}\right)$$

Dividing both sides by $\sqrt{x^2 + y^2}$, we obtain

$$1 = \left(\frac{y}{x^2}\right)$$

or
$$y = x^2$$

which is the equation we set out to find.

EXAMPLE: Change the following equation to an equation in rectangular coordinates.

$$\rho = \frac{3}{\sin \theta - 3 \cos \theta}$$

SOLUTION: Written without a denominator the polar equation is

$$\rho \sin \theta - 3 \rho \cos \theta = 3$$

Using the transformations

$$\rho \sin \theta = y$$

$$\rho \cos \theta = x$$

we have $\quad y - 3x = 3$

as the equation in rectangular coordinates.

Practice Problems

Change the equation in problems 1 through 4 to equations having polar coordinates.

1. $x^2 + y^2 = 4$

2. $\left(x^2 + y^2\right) = a^3 x^2$

3. $3y - 7x = 10$

4. $y = 2x - 3$

Change the equations in the following problems to equations having Cartesian coordinates.

    5. $\rho = 4 \sin \theta$

    6. $\rho = \sin \theta + \cos \theta$

    7. $\rho = a^2$

Answers

    1. $\rho = \pm 2$

    2. $\rho = a \sqrt{a \cos \theta}$

    3. $\rho = \dfrac{10}{3 \sin \theta - 7 \cos \theta}$

    4. $\rho = \dfrac{-3}{\sin \theta - 2 \cos \theta}$

    5. $x^2 + y^2 - 4y = 0$

    6. $x^2 + y^2 = y + x$

    7. $x^2 + y^2 = a^4$